Science Confirms the Existence of God

Kevin Bradford Ornellas

Copyright

ISBN: 978-1-7354667-3-6

Publisher: Kevin Bradford Ornellas

Honolulu, HI 96816

Cover illustrations were done by leraynne on Fiverr (https://www.fiverr.com/leraynne)

Printed in the United States of America

Dedication

Glory and honor to God and to the Lord Jesus Christ.

Honor to Eugene Ornellas and Linda Ornellas.

Dedicated to Mrs. Sue Wankowicz and Mr. Paul G. Wankowicz for your inspiration and very positive influence on my life.

Table of Contents

Preface

Nearly every chapter of this book makes at least one direct reference to the Genesis Creation Account in Genesis chapters 1 and 2. I will include the main text of Genesis chapter 1 and the beginning of Genesis chapter 2 from the Bible translation that I used most heavily throughout the book, Young's Literal Translation (1898).

To refresh the reader's memory, here is chapter 1 and the beginning of chapter 2 of the Book of Genesis[1], the Biblical Genesis Creation Account.

Young's Literal Translation (1898)

Chapter 1

1 In the beginning of God's preparing the heavens and the earth--

2 the earth hath existed waste and void, and darkness is on the face of the deep, and the Spirit of God fluttering on the face of the waters,

3 and God saith, `Let light be;' and light is.

4 And God seeth the light that it is good, and God separateth between the light and the darkness,

5 and God calleth to the light `Day,' and to the darkness He hath called `Night;' and there is an evening, and there is a morning--day one.

6 And God saith, `Let an expanse be in the midst of the waters, and let it be separating between waters and waters.'

7 And God maketh the expanse, and it separateth between the waters which are under the expanse, and the waters which are above the expanse: and it is so.

8 And God calleth to the expanse `Heavens;' and there is an evening, and there is a morning--day second.

9 And God saith, `Let the waters under the heavens be collected unto one place, and let the dry land be seen:' and it is so.

10 And God calleth to the dry land `Earth,' and to the collection of the waters He hath called `Seas;' and God seeth that it is good.

11 And God saith, `Let the earth yield tender grass, herb sowing seed, fruit-tree (whose seed is in itself) making fruit after its kind, on the earth:' and it is so.

12 And the earth bringeth forth tender grass, herb sowing seed after its kind, and tree making fruit (whose seed is in itself) after its kind; and God seeth that it is good;

13 and there is an evening, and there is a morning--day third.

14 And God saith, `Let luminaries be in the expanse of the heavens, to make a separation between the day and the night, then they have been for signs, and for seasons, and for days and years,

15 and they have been for luminaries in the expanse of the heavens to give light upon the earth:' and it is so.

16 And God maketh the two great luminaries, the great luminary for the rule of the day, and the small luminary--and the stars--for the rule of the night;

17 and God giveth them in the expanse of the heavens to give light upon the earth,

18 and to rule over day and over night, and to make a separation between the light and the darkness; and God seeth that it is good;

19 and there is an evening, and there is a morning--day fourth.

20 And God saith, `Let the waters teem with the teeming living creature, and fowl let fly on the earth on the face of the expanse of the heavens.'

21 And God prepareth the great monsters, and every living creature that is creeping, which the waters have teemed with, after their kind, and every fowl with wing, after its kind, and God seeth that it is good.

22 And God blesseth them, saying, `Be fruitful, and multiply, and fill the waters in the seas, and the fowl let multiply in the earth:'

23 and there is an evening, and there is a morning--day fifth.

24 And God saith, `Let the earth bring forth the living creature after its kind, cattle and creeping

thing, and beast of the earth after its kind:' and it is so.

25 And God maketh the beast of the earth after its kind, and the cattle after their kind, and every creeping thing of the ground after its kind, and God seeth that it is good.

26 And God saith, `Let Us make man in Our image, according to Our likeness, and let them rule over fish of the sea, and over fowl of the heavens, and over cattle, and over all the earth, and over every creeping thing that is creeping on the earth.'

27 And God prepareth the man in His image; in the image of God He prepared him, a male and a female He prepared them.

28 And God blesseth them, and God saith to them, `Be fruitful, and multiply, and fill the earth, and subdue it, and rule over fish of the sea, and over fowl of the heavens, and over every living thing that is creeping upon the earth.'

29 And God saith, `Lo, I have given to you every herb sowing seed, which is upon the face of all the earth, and every tree in which is the fruit of a tree sowing seed, to you it is for food;

30 and to every beast of the earth, and to every fowl of the heavens, and to every creeping thing on the earth, in which is breath of life, every green herb is for food:' and it is so.

31 And God seeth all that He hath done, and lo, very good; and there is an evening, and there is a morning--day the sixth.

Chapter 2

1 And the heavens and the earth are completed, and all their host;

2 and God completeth by the seventh day His work which He hath made, and ceaseth by the seventh day from all His work which He hath made.

3 And God blesseth the seventh day, and sanctifieth it, for in it He hath ceased from all His work which God had prepared for making.

The other critical passage in the Bible that I use to be able to calculate the time frame for the Genesis Creation Account throughout this book is 2Peter 3:8. Here is 2Peter 3:8 in the New Heart English Bible's translation of the verse[2]:

But do not forget this one thing, beloved, that one day is with the Lord as a thousand years, and a thousand years as one day. (2Peter 3:8, NHEB)

---

[1] I've changed the spacing within the passage to improve readability by having the verse numbers on the far left. Each new Day in Genesis is indented by several spaces as well for the reader's convenience.

[2] The New Heart English Bible from Wayne A. Mitchell is in the public domain.

Chapter 1: Introduction

This book demonstrates that God exists by using the secular data of Modern Science and the secular discipline of Mathematics. I develop a simple mathematical model based primarily on the statements made by two Biblical individuals: the Genesis Writer and Saint Peter. Insight into the meaning of 2Peter 3:8 is needed to design this book's mathematical model. Using this model, I found that there is a direct correlation between the times and order of events predicted by the Bible and the times and order of ancient events revealed by modern scientific research. Since the correlation is direct, the weight of the evidence forms a true and fair basis for confirming the factual existence of God based on the rational grounds of Modern Science.

Chapter 2: Overview of the Line of Reasoning for the Statement 'Science Confirms the Existence of God'

The line of reasoning for the statement 'Science confirms the existence of God' requires generating a 'model'. This mathematical model is founded on statements from Scripture. This model provides measurable time-windows against which the timings of ancient scientific events can be compared and tested. This book's model is successful in showing the very direct correlation of the Genesis seven day creation account with scientifically measured dates of the first occurrence of ancient events in the history of our solar system.

Insight into the meaning of 2Peter 3:8 is needed to design this book's mathematical model. The insight derived from the meaning of 2Peter 3:8 validly assigns a 365-million-year length of time for one unit of the Lord's Time in the Genesis Creation Account. By using the 365 million year length of time for one unit of the Lord's Time in the Genesis Creation Account, the first known occurrence for each of 13 chosen events described in the Genesis Creation Account correlates in an exacting fashion with the scientifically determined timeline for the same event in the history of our solar system. This book compares the timing and order of 13 events described in the Genesis Creation Account to the timing and ordering of the same 13 events by modern scientific means.

Now I will cover in more detail the model design that allows this chain of reasoning. The Genesis Creation Account subdivides into 7 Nights and 7 Days with the Night going first: '…there is an evening, and there is a morning…'.[3] So there are 2 units of time, 1 for the Night and 1 for the Day totally 14 units of the Lord's Time in the seven day Genesis Creation Account. Each unit of time is 365 million years long, thus the total length of the Genesis Creation Account is 5,110 million years (5.11 billion years). 5.11 billion years is the result of 14 units x 365 million years per unit.

An extensive review of scientific literature was done for researching this book, which led to identifying the scientific timings of 13 events described in the Genesis Creation Account. The scientifically determined first known occurrence of ancient events in the history of our solar system that directly correlate to the Biblical Genesis Creation Account include the following thirteen events: the Earth, the Sun, the Sky, the Oceans, the Dry Land, the Green, Seed, the Seasons, the Teeming Living Creature as a Marine Protozoan, the Living Creature that Crawls (in the sea) as a Sea Sponge, the Flying Creatures as Airborne Protozoa, Animal of the Earth as an Earthworm, and Land Plants. The model designed in this book demonstrates that the known scientific data line up with the Genesis Creation Account for both the timing of the events and the order of the events.

Let's look at the critical verse from which we can validly derive 365 million years of our time for 1 unit of the Lord's Time.

But do not forget this one thing, beloved, that one day is with the Lord as a thousand years, and a thousand years as one day. (2Peter 3:8, NHEB)

Below is 2Peter 3:8 with my notes in parentheses to improve clarity:

But do not forget this one thing, beloved, that one day is with the Lord as a thousand years (in Heaven), and a thousand years as one day (in Heaven). (2Peter 3:8, NHEB)

There are three timescales in play in 2Peter 3:8: the Lord's Time, our time, and Heaven Time!

With 3 timescales in play, a mathematical solution can be found to the question of "What is the length of one day of the Lord's Time?" 1 Day with the Lord is 1000 years in Heaven. We convert years in Heaven to days in Heaven. Multiplying the number, 1000 years in Heaven, by 365 because there are 365 days in a year, makes 365,000 Days in Heaven. Next, we convert Days in Heaven Time to our time by multiplying by 1000, because there are 1000 years of our time for each day in Heaven. The overall multiplication is 1000 x 365 x 1000 = 365,000,000 years of our time. 365,000,000 years of our time is one day of the Lord's Time. The logic and the mathematical calculations to demonstrate the statement that 1 Day (or separately, 1 Night) of the Lord's Time is 365,000,000 years of our time will be fully discussed in Chapter 6.

The exercise of measuring how supportive the model is for determining the sequence and timing of the scientific events within the history of the solar system is important. Also important is that this exercise gives us the ability to measure the probability of divine inspiration occurring for the Genesis Writer and St. Peter from a skeptical point-of-view. The correct statements regarding the timing and ordering of the 13 creation events, later substantiated by scientific data, would indicate divine inspiration. Divine inspiration would and does validate the existence of the Divine i.e. God.

Atheism -- as a school of thought -- maintains that there are exactly zero divine entities – that there is no God. Atheism is restricted to exactly one narrative: that all the Biblical writers' correct statements involved in the Genesis Creation Account were 'guesses in the dark' and that they got 'lucky' by chance. Atheism must make this case, because if any Biblical writer had divine insight, divine insight would imply the existence of the Divine, the source of the knowledge.

Under the rules of a skeptical and reasonable atheistic mindset, we can still derive a weighing of probabilities that demonstrate that God's existence is very probable despite that mindset. By measuring the likelihood of the Biblical events correlating with scientific events under chance-only conditions, we can obtain a measure for the probability of either God's existence or non-existence under a skeptical-and-reasonable atheistic mindset.

The chance of having the relative order of the Genesis Creation Account events correlate correctly with regards to the observed scientific evidence is small. There are 13 factorial possible arrangements for 13 events, with only 1 arrangement able to match the correct arrangement for the relative order of the scientific data. But the Genesis Writer did provide the correct arrangement – the correct relative order for the 13 events. The chance of this occurring is one over 13 factorial = 1/(13!) = 1 / 6,227,020,800: about a 1-in-a-6 billion shot.

We're not done yet. The chance of having the exact times of the Genesis Creation Account events correlate with the observed scientific evidence is even smaller. Since there are 14 subdivisions of time in the Genesis Timeline – 7 Nights and 7 Days – there is no greater than a 1-in-14 shot for the scientific evidence for one ancient event's timing to fall on the Day that the Biblical writers required the event to fall on (and not on the Night). These 13 events correlate with the appropriate Day time-window within the 14 time-windows. With 14 time-windows used in the probability calculation, the probability is $(1/14)^{13} = 1 / 793,714,773,254,144$, which is less than a 1 in 793 trillion shot of the timings of 13 ancient events correlating. But they did; the Biblical authors did give the correct placement of the scientific data into the correct timeslot.

Atheism, which has to make the argument that the timings and relative order of the events were correct only by chance, has less than a 1 in 4 septillion chance of correctness. A chance of less than a 1 in 4 septillion is the result of multiplying the chance of anyone correctly guessing, in the absence of information, the timings as well as the order of 13 events. The probability of getting 13 events in the correct relative order and falling in the correct time-slots is the probability of the relative order times the probability of the correct day placement. This number is $(1/6,227,020,800) \times (1/14)^{13} =$ $1 / 4,942,478,402,320,838,374,195,200$, which less than 1 in 4 septillion chance that the order and timings of the 13 events could be guessed correctly. Atheism gets 'stuck' with this number.

Since God either exists or does not exist, and the certainty of God's non-existence estimated under skeptical conditions can be shown to be no better than the very small number of 1 / 4,942,478,402,320,838,374,195,200, this evidence is strong enough on its own to conclude that God exists. Since the scientific events correlate with the Biblical Creation Account, Atheism's best-case scenario still implies that God's existence is at least 99.9999999999999999999 % certain.

Thus, the statement that 'Science confirms the Existence of God' is true, reasonable, and fair.

Table 1 is a summary of most of the information presented in this book. In Table 1, the first column lists the Day or Night in Genesis. The second column uses the model designed in this book to provide the calculated timeline of 2Peter3:8. The third column states the thirteen Biblical Account Reference Events chosen to be examined. The fourth column states the determination of time for each Reference Event being examined using modern scientific evidence for determining the first known occurrence in the history of our solar system. The Genesis Creation Account goes from the Beginning of Night 1 to the end of Day 7.

TABLE 1: The Genesis Timeline, the Calculated Timeline Based on 2Peter 3:8, the Biblical Account's Reference Events, and the Scientific Determination of Time for Reference Events.			
Timeline in Genesis 1 & 2	**Calculated Timeline Based on 2Peter 3:8 (years ago)**	**Biblical Account Reference Event**	**Scientific Determination of Time for Reference Event (plus citations)**
Beginning of Night One	5,110,000,000 (i.e. 5.11 billion years ago)		
Beginning of Day One	4,745,000,000	'the earth'	The earth formed at about or slightly before 4.65 billion years ago.[4]
		'the deep' and 'the face of the waters'	Space ice supply predating the Sun[5]
		'the light'	The Sun forms at around 4.5682 billion years ago.[6]
Beginning of Night Two	4,380,000,000		
Beginning of Day Two	4,015,000,000	'the sky'	Evidence exists for the atmosphere since the end of the Hadean Eon[7], which is about 4.0 billion years ago[8]. There is no evidence for an 'early' atmosphere.[9]
		'God…divided the waters which were under the expanse from the waters which were above the expanse'	The hydrosphere existed by approximately 3.85 billion years ago[10], and there is no direct evidence for a liquid water-based hydrosphere timing before 4.0 billion years ago.[11]
Beginning of	3,650,000,000		

Night Three			
Beginning of Day Three	3,285,000,000	'let the dry land be seen'	First landmass to appear above the waterline occurred at about 3.1 billion years ago[12]
		'green'	Divergence within land-colonizing bacteria between Cyanobacteria 'blue-green algae' and Chloroflexia estimated at 3.069 billion years ago (CI 2.922 to 3.184 billion years)[13]
		'herb sowing seed'	Estimated timing for divergence of endospore-forming bacteria phylum Firmicutes is about 3.054 billion years ago.[14]
Beginning of Night Four	2,920,000,000		
Beginning of Day Four	2,555,000,000	'and for seasons'	Evidence of strong seasonality from 2.4 billion to 2.3 billion years ago during the Huronian Glaciation period[15]
Beginning of Night Five	2,190,000,000		
Beginning of Day Five	1,825,000,000	'Let the waters teem with the teeming living creature'	Modern eukaryotes emerged late in Earth history (<1.84 Ga) [16], and late Palaeoproterozoic rocks preserve

			evidence for a moderate diversity of eukaryotic organisms.[17]
		'living creature that crawls, with which the water swarms, according to their species'	Evidence of chemical that is diagnostic for demosponges (a type of sea sponge) preserved in 1.8 billion year old sediments.[18]
		'every winged flying creature, according to their species'	With multiple amoebas noted as being present in atmospheric air[19], the grouping that contains amoebas, Amoebozoa, is estimated to have come into existence is 1.538 billion years ago.[20]
Beginning of Night Six	1,460,000,000		
Beginning of Day Six	1,095,000,000	'beast of the earth'	Worm burrows in sandstone in India that are from more than 1.0 billion years old up to as early as 1.1 billion years ago.[21]
		'every tree in which is the fruit of a tree sowing seed'	Scientific analysis supported that crown embryophytes originated sometime between 980 and 682 million years ago.[22]
Beginning of Night Seven	730,000,000		The newest known galaxy according to NASA[23], which formed about 500 million years ago.
Beginning of Day Seven	365,000,000	'the heavens…complet	No known galaxy has formed within the last

		ed, ...and God...ceaseth by the seventh day from all His work which He hath made'	365 million years.
End of Day Seven	0		

Note that all of the creation events in our solar system occur in the Daytime slots. No creation events in our solar system had the weight of their evidence occurring in the Nighttime slots.

Scientific support for the information in Table 1 is extensive, spanning many disciplines: Astronomy, Geology, Chemistry, Biology, Paleontology, and Botany, among others.

It is important to understand that the Biblical writers wrote or had their words recorded before the advent of modern science. The Genesis Writer wrote before Saint Peter. St. Peter lived over a thousand years before the time of modern scientific discoveries made by using scientific means for the timings of Creation events.

As to the question of whether the Genesis Writer and Peter were divinely inspired or just guessed the order and timing of the Genesis Creation Events, critics of purportedly divinely-inspired people would say that the 'divinely inspired' people would have come by their knowledge by 'reverse engineering' their statements to the data. This is not possible for the Genesis Writer and St. Peter because the relevant scientific data was not available until more than a thousand years after the Genesis Writer and Saint Peter lived.

The chance of supplying the correct order and correct times without prior knowledge for thirteen events in the Genesis Creation Account is absurdly small. The probability will be discussed in greater detail in Chapter 31.

If the Genesis Writer and St. Peter spoke in accordance with divine inspiration, the truth of their statements would be certain. After reading this book, the reader will likely conclude that the Genesis Writer and Saint Peter gave the correct order and the correct times for the ancient events.

The Genesis Writer and St. Peter, because they spoke with divine inspiration, validate the existence of the Divine i.e. God.

All this follows from a proper understanding of what Scripture asserted, what the associated math implies, and what Science has now revealed.

Chapter 3: Book's Contents

Now for an outline of the book's contents. Chapter Four covers an overview of the Biblical Genesis Creation Account's timeline: a model that can be generated from Scripture. Chapters Five thru Nine cover establishing the design of the model and justifying the numbers for the model. Chapter Five covers that the Biblical text implies 7 Nights and 7 Days along with justification for there being 14 units. Chapter Six, a very important chapter, covers how I validly extract a time length of 365,000,000 years for one (1) day of the Lord's Time from Scripture. Chapter Seven covers the total time length of the Genesis Creation Account being 5.11 billion years. Chapter Eight covers the starting and ending times for the model. Chapter Nine covers the justification for the first unit in the sequence being a Night and that none of the creation events related to Earth and our solar system can occur during the Nights. Chapter Ten showcases the timeline that the model results in.

With the model and its timeline established, Chapter Eleven thru Chapter Twenty-Eight cover the scientific first occurrences of the 13 chosen events. Most of these chapters are 'scientific event chapters' and show how an individual scientific event relates to the model's timeline and its testable time-windows. A summary of the updated information in a table format appears near the end of most chapters.

Chapter Twenty-Nine covers Day Seven and the knowledge that no known galaxies have been formed during Day Seven (i.e. 365,000,000 years ago to the present).

Chapter Thirty covers the probability that the Genesis Writer would correctly order the thirteen events that I cover in Day One thru Day Six.

Chapter Thirty-One covers the probability that the Genesis Writer and Saint Peter would provide the correct Day for each of the 13 Day One through Day Six individual events chosen for this book.

Chapter Thirty-Two covers the probability of both the Genesis Writer and of Saint Peter - the Biblical authors in question - taking a guess and being correct by chance. Chapter Thirty-Two also provides the understanding and reasoning to arrive at a value for the probability of the existence of the Biblical God if the galaxy statement discussed in Chapters Twenty-Nine and

Thirty-Three is not taken into account. Chapter Thirty-Three discusses and examines the chance of the occurrence of no new galaxies formed on Day Seven, although it is unquantified because I do not have access to all of NASA's age-related galaxy data. Chapter Thirty-Four covers the best-case probability for Atheism as well as the overall worst-case probability for the existence of God. Chapter Thirty-Five covers the minimum probability for the existence of God under reasonable but skeptical assumptions.

The Conclusion is that Science confirms the existence of God.

Following the Conclusion are the Acknowledgments - thank you so much everyone! The Appendices then follow, which cover information that is not necessary but may be of interest to the reader. Appendix A covers the Julian Calendar system versus the Gregorian Calendar system. Appendix B provides supplementary figures, pictures, and images. Appendix C includes a statement from NASA's blog about the number of galaxies in the observable universe.

Following the Appendices are the References along with the Endnotes.

I hope you enjoy this book.

Chapter 4: Overview of the Model for the Biblical Timeline

Six questions are essential for building a mathematical model for the Genesis Creation Account.

Q1. How many units are there within the model?

Q2. How long is each unit within the model?

Q3. How long is the total time for the model?

Q4. When does the model start?

Q5. When does the model end?

Q6. Is the first unit a Day or a Night?

I will now answer each question.

Q1. How many units are there within the model?

Answer: There are 14 units within the model. 1 unit for each Day, and 1 unit for each Night. The Genesis Creation Account is 7 Days and 7 Nights.

Q2. How long is each unit within the model?

Answer: Each Day in the Genesis Creation Account is 365,000,000 Julian years, and each Night is 365,000,000 years as well. So each unit is 365,000,000 Julian years long.

Q3. How long is the total time for the model?

Answer: 5.11 billion years.

Q4. When does the model start?

Answer: The model started approximately 5.11 billion years ago.

Q5. When does the model end?

Answer: I treat the Biblical model as ending at Adam and Eve's exit from the Garden of Eden, which is estimated to have occurred approximately 6,000 to 10,000 years ago. 6,000 to 10,000 years ago until now is an insignificant timespan relative to the model's total timespan of 5.11 billion years. I treat approximately 10,000 years ago as "0" time.

Q6. Is the first unit a Day or a Night?

Answer: The first unit is a Night. The sequence goes Night, Day, Night, Day, etc. The text and other knowledge support this statement.

I will now summarize the model. The model consists of 14 units of time: one unit for each of the 7 Days, and one unit for each of the 7 Nights. Each unit is 365,000,000 Julian years long. The total time of the model is 5.11 billion years exactly. The model begins 5.11 billion years ago. I treat the model as ending at Adam and Eve's exit from the Garden Eden 6,000 to 10,000 years ago. The model starts with a Night, so Night One goes first, then Day One, then Night Two, then Day Two, etc.

I will spend Chapters 5 thru 9 building this model for the Genesis Creation Account.

Chapter 5: Building the Model — the Model's Number of Units

Regarding Question #1 from Chapter Four – "How many units are there within the model?" – there are 14 units within the model. In Genesis Chapter One and spanning into the beginning of Genesis Chapter Two, the text implies 7 Nights and 7 Days – a total of 14 units.

Does the Genesis Creation Account contain 7 units of time or 14 units of time? The text is clear that there is evening, and that there is morning, and that the Lord draws a distinction between Day and Night, and separates the two in Genesis chapter 1.

And God seeth the light that it is good, and God separateth between the light and the darkness, and God calleth to the light `Day,' and to the darkness He hath called `Night;' and there is an evening, and there is a morning--day one. (Genesis 1:4-5, YLT)

So Day and Night are distinct according to God. Consider the beginning of the prayer of Moses, which is the one psalm we have from Moses in the Bible, Psalm 90.

Psalm 90

A prayer of Moses, the man of God.

O Lord, you have been our protector through all generations.
Even before the mountains came into existence,
or you brought the world into being,
you were the eternal God.
You make mankind return to the dust,
and say, "Return, O people."

Yes, in your eyes a thousand years
are like yesterday that quickly passes,
or like one of the divisions of the nighttime.

(Psalm 90:0-4, the New English Bible)

Moses referred to the Lord with the following statement: "…in your eyes a thousand years are like yesterday that quickly passes, or like one of the divisions of the nighttime."

So 1000 years are like a day 'yesterday', or like one of the divisions of the nighttime, which could be a reference to a Night.

Given the above, the Lord draws a distinction between Day and Night. So there are 7 Days and 7 Nights implied in the Genesis Creation account. The following Math equation can formalize this understanding.

Total time in the Genesis Creation Account = 7 Days of the Lord's time + 7 Nights of the Lord's time

God draws a distinction between day and night in Genesis. Jesus also uses a 12-hour daytime in His teaching. Here is the statement by Jesus to this effect from John's Gospel:

Jesus answered, `Are there not twelve hours in the day? if any one may walk in the day, he doth not stumble, because the light of this world he doth see; and if any one may walk in the night, he stumbleth, because the light is not in him.' (John 11: 9-10, YLT)

Jesus considers there to be 12 hours in a day. Jesus assigns the Day as 12 hours. Our modern concept is to group day and night together for a time for a date, but Jesus's testimony here can be viewed as the 12 daylight hours form the Day. So what about the Night? Assigning 12 hours to the Day should suggest that the night includes the remaining 12 hours to make the total 24 hours. So 1 Day of the Lord's Time and 1 Night of the Lord's time are of equal length in terms of time; they are each 12 hours of the Lord's Time apiece. I note this equality below.

1 Day of the Lord's time for time length = 1 Night of the Lord's time for a time length

We used Jesus's 12-hour-day statement, which implied that the length of 1 of the Nights is equal to the length of 1 of the Days. 7 days + 7 nights of Lord's Time is equal to 14 units of Lord's Time in terms of time length.

So the answer to Question #1 from Chapter Three, "How many units are there within the model?" is that there are fourteen units. One unit for each Day in the Genesis Creation Account, and one unit for each Night in the Genesis Creation Account. The Genesis Creation Account is 7 Days and 7 Nights long.

Chapter 6: Building the Model — Defining the Length of each Unit of Time

Regarding Question #2 from Chapter Four -- "How long is each unit within the model?" -- there are two units of the Lord's Time per 24-hour period. That is 1 unit for the Night and 1 unit for the Daytime. Each unit of the Lord's Time is 365,000,000 Julian years long. Thus, there are 365,000,000 Julian years for 1 Day of the Lord's Time, which accounts for the 12-hour period of Daytime that is being used in the Genesis Creation Account. Each night in the Genesis Creation Account is also 365,000,000 Julian years. Typically, scientists use Julian years for measuring the time for ancient events.

To find out how long each unit of the Lord's Time is, we need to understand a critical verse - 2Peter 3:8.

"But do not forget this one thing, beloved, that one day is with the Lord as a thousand years, and a thousand years as one day." (2Peter 3:8, NHEB)

St. Peter's kind demand is "But do not forget" from the reader.

This critical sentence has historically been misinterpreted. Misinterpretation of this sentence has clouded the truth of this crucial message for Mankind. Understanding the concept of this Biblical passage is key to understanding the model and a very important insight in this book.

Let's focus on the message that Saint Peter conveys:

"...that one day with the Lord is as a thousand years, and

a thousand years as one day."

The mathematically sound interpretation for the critical verse 2Peter 3:8 relies on the sentence's end, more specifically that 'one day.' refers to one day of the first-mentioned 'thousand years', and not to the initial 'one day with the Lord'.

We can break down Saint Peter's statement in this way:

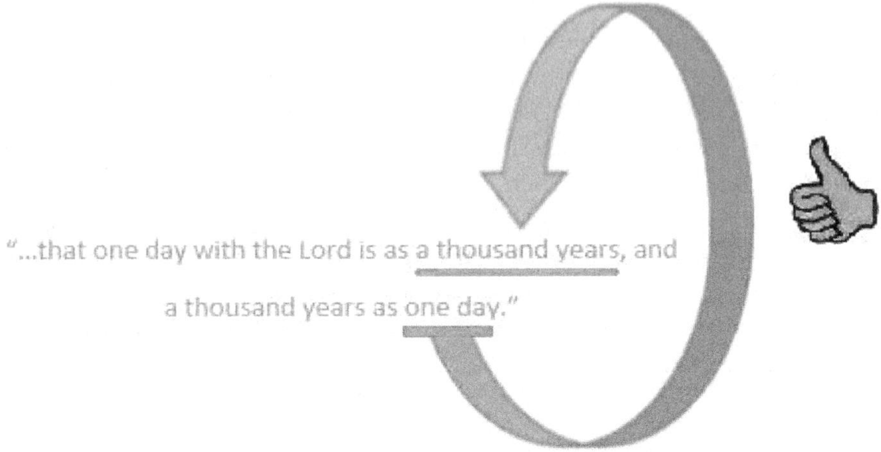

Follow the arrow in the above picture. The arrow points to what I believe Saint Peter was referring to. I will show this to be the case by providing an interpretation of the text not generally made known in the modern era by other authors.

A key insight in this chapter and for the rest of the book is that this reference implies composition rather than mere repetition of the earlier portion of the sentence. The end of the sentence, 'one day', is referring to 'one day' of the first aforementioned 'a thousand years'. In other words, that 'one day' means 'one day' of the same timescale of the first-mentioned thousand years. This reasoning is like saying: 'one day' of <u>those years</u> (the first mention of 'a thousand years').

The fourth time referring to the second time in the sentence implies that 3 timescales are in play.

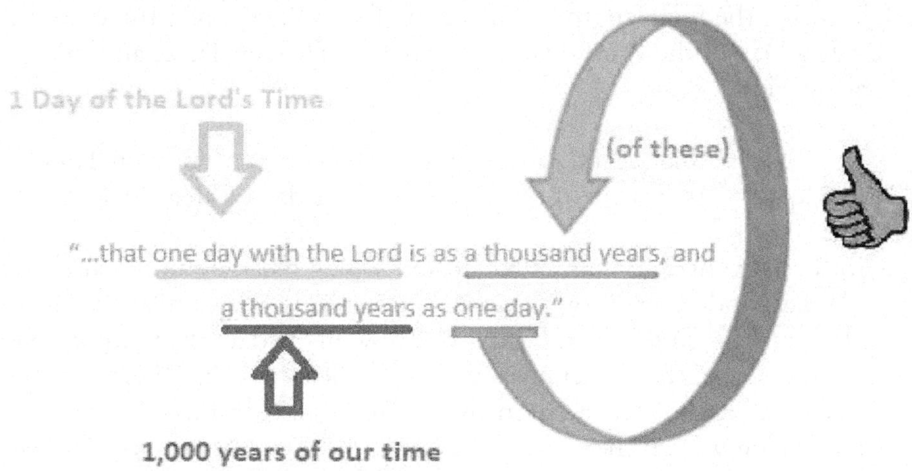

In 2Peter 3:5, Saint Peter warns that there would come a time where scoffers would '…willingly forget, that there were heavens from of old' (2Peter 3:5, NHEB). Let us not make the same mistake as the scoffers by forgetting Heaven; Heaven Time needs to be included in our calculation to understand the correctness of the 2Peter 3:8 verse. Let's see where remembering Heaven and including its timescale leads us.

Saint Peter mentioned four (4) increments of time in 2Peter 3:8. He talked about 4 distinct amounts of time for a reason -- to provide the correct length for 1 day of the Lord's time in years. The first unit is 'a Day of the Lord's Time.' The second unit is 'Years in Heaven Time.' The third unit is 'years', which was understandable to people while Peter was alive in the Roman Empire era and is also understandable to us today. The fourth unit is in 'Days of Heaven Time.'

If notes were written to aid the reader in interpreting the verse, it could look something like this: St. Peter with commentary, "But do not forget this one thing, beloved, that one day is with the Lord as a thousand years (in Heaven), and a thousand years (in human time) as one day (in Heaven)." (2Peter 3:8, New Heart English Bible)

Now let's find the measure of time for one (1) day of the Lord's time in human years. That is the measure that we will use for the length of a Day in the Genesis Creation Account. We know the length of 1,000 years. With that said, knowing the time length of one of the four values leaves three unknown values: 1) a Day of the Lord's Time; 2) a year of Heaven Time, and 3) a day of Heaven Time.

(Eq. 1): 1 day of the Lord's time = 1000 years of Heaven Time
(Eq. 2): 1000 years of our time = 1 day of Heaven Time

Saint Peter wanted to communicate something meaningful. He used the terms "a day" and "years" for a reason. These terms carry a relationship with them. If someone on the street asked you how many days there are in a year, you would respond that generally -- except for 'leap years': "There are 365 days in a year." Right? The relationship of days to a year implies another equation, which we can then use: there being 365 Days of Heaven Time in a year of Heaven Time. This allows us to retrieve and justify the answer to this key question: "How long is 1 Day of the Lord's Time?"

So here is the third equation.

(Eq. 3): 1 year of Heaven Time = 365 days of Heaven Time

Now we have 3 equations. With 3 equations and 3 unknown constant values, we can solve for the value of each previously unknown constant value, including the length of our human time for 1 Day of the Lord's time.

Equation 1:	**1 day of the Lord's Time**	**=**	**1000 years of Heaven Time**
Equation 2:	**1000 years of our time**	**=**	**1 day of Heaven Time**
Equation 3:	**1 year of Heaven Time**	**=**	**365 days of Heaven Time**

Here is the calculation displayed visually:

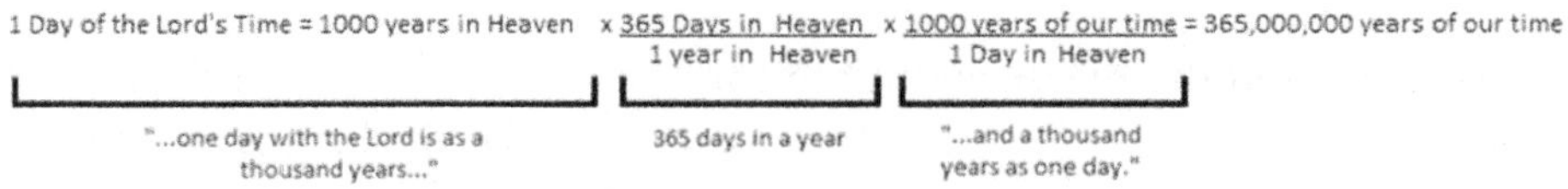

Here is the algebra with a short explanation at each step.

(Eq. #1) 1 day of the Lord's Time = 1000 years of Heaven Time

(Eq. #2) 1000 years = 1 day of Heaven Time

(Eq. #3) 1 year of Heaven Time = 365 days of Heaven Time

$$\frac{\text{1 day of the Lord's Time}}{\text{1000 years of Heaven Time}} = \frac{\text{1000 years of Heaven Time}}{\text{1000 years of Heaven Time}} = 1$$

Dividing Both Sides of Eq. #1 by its Right-Hand-Side

$$\frac{\text{1000 years}}{\text{1000 years}} = \frac{\text{1 day of Heaven Time}}{\text{1000 years}} = 1$$

Dividing Both Sides of Eq. #2 by its Left-Hand-Side

$$\frac{\text{1 year of Heaven Time}}{\text{365 days of Heaven Time}} = \frac{\text{365 days of Heaven Time}}{\text{365 days of Heaven Time}} = 1$$

Dividing Both Sides of Eq. #3 by its Right-Hand-Side

$$1 = 1 \times 1 \times 1 = \frac{\text{1 day of the Lord's Time}}{\text{1000 years of Heaven Time}} \times \frac{\text{1 year of Heaven Time}}{\text{365 days of Heaven Time}} \times \frac{\text{1 day of Heaven Time}}{\text{1000 years}}$$

From Eq. #1 ⇧ From Eq. #3 ⇧ From Eq. #2 ⇧

Substitution

$$1 = \frac{\text{1 day of the Lord's Time}}{\text{1000 } \cancel{\text{years of Heaven Time}}} \times \frac{\text{1 } \cancel{\text{year of Heaven Time}}}{\text{365 } \cancel{\text{days of Heaven Time}}} \times \frac{\text{1 } \cancel{\text{day of Heaven Time}}}{\text{1000 years}}$$

Cancel Units

$$1 = \frac{\text{1 day of the Lord's Time}}{1000} \times \frac{1}{365} \times \frac{1}{\text{1000 years}}$$

Remove Cancelled Units

$$1 = \frac{\text{1 day of the Lord's Time}}{\text{365,000,000 years}}$$

Perform two multiplications

$$\text{365,000,000 years} \times 1 = \frac{\text{1 day of the Lord's Time}}{\text{365,000,000 years}} \times \text{365,000,000 years}$$

Multiply both sides by the denominator

$$\text{365,000,000 years} = \text{1 day of the Lord's Time}$$

Cancel and Reduce Terms

$$\text{1 day of the Lord's Time} = \text{365,000,000 years}$$

Switch Left-Hand-Side with the Right-Hand-Side

Sometimes, it's easier to view the equations in terms of math variables. We can solve this in another way to make the outcome clearer:

X: 1 day of the Lord's Time
Y: 1 year of Heaven Time
Z: 1 day of Heaven Time

Find X for:
(Eq.1) 1 X = 1000 Y

(Eq.2) 1000 = 1 Z
(Eq.3) 1 Y = 365 Z
Where the value is in human years of our time.
Note that * means multiplication.

Start with Equation 2: 1000 = 1 Z
Substitute in the Value for Z from Equation 2 into Equation 3: 1 Y = 365 * (1000)
Simplify to solve for 1 Y: 1 Y = 365,000
Substitute in the value for Y into Equation 1: 1 X = 1000 * (365,000)
Simplify to solve for 1 X: 1 X = 365,000,000

And there we have it, X, which is 1 day of the Lord's Time is 365,000,000 years long in human time.

So, at this point, you might be thinking: "Wait, what? I thought the verse just meant 1 Day with the Lord is 1000 years of our time." That interpretation has a flawed built-in assumption. Specifically, the flawed assumption is that the first mentioned 'one day with the Lord' and the fourth mentioned 'one day' are the same. However, you might be unintentionally inserting a word that is not there – as there is no 'with the Lord' marking the fourth mentioned 'one day'. It's not in most English language Bibles, and more importantly, it's not in the old manuscripts that I examined. The ancient Greek word for 'Lord' or 'Master' is 'kurios' which occurs in 2Peter3:8 only once in the verse, not twice in the verse. Here are the names of three old manuscripts in which 'kurios' occurs only once:

1) the Textus Receptus of the King James Version;
2) the Byzantine Text;
3) the Elzevir 'Textus Receptus.'

During modern times, those trying to interpret the verse have often made the mistake of assuming that the fourth time refers to the first time in spite of the fourth time not being marked by the term 'with the Lord'. This oversimplification of Saint Peter's statements quite frankly results in a timeline that is inconsistent with modern scientific findings. This mistake of oversimplifying this statement has led many people to extract the incorrect time for what the Bible was stating. A timing of 365,000,000 years for one Day of the Lord's Time can be validly extracted from the text of 2Peter 3:8. We will see in the next chapter how this validly extracted time fits into the model and how it determines the model's overall time for the Genesis Creation Account.

Chapter 7: Building the Model — the Model's Overall Time

This chapter justifies the answer to question #3 from Chapter Four – "How long is the total time for the model?"

We recall that there are 7 Days and 7 Nights for the Genesis Creation Account and that a Day of the Lord's Time is the same length of time as a Night of the Lord's Time, which was justified in Chapter Six. The result was that 7 Nights and 7 Days of the Lord's Time are equal to 14 units of the Lord's Time in terms of time length.

We will substitute these 14 units of the Lord's Time for the Total Number of Days and Nights within the Genesis Creation Account. We will substitute in 365,000,000 years for one Day of the Lord's time, which we justified in Chapter Six. The calculation for the total time for the Genesis Creation Account is as follows:

Total Time for the Genesis Creation Account = 14 x 365,000,000 years = 5,110,000,000 years

Total Time for the Genesis Creation Account = 5,110,000,000 years

The total time for this book's model of the Genesis Creation Account is 5,110,000,000 years, 5.11 billion years.

Chapter 8: Building the Model — the Starting Time and the Ending Time of the Model

This chapter will discuss the answers that I gave to question #4 and question #5 from Chapter Four.

Question #4: When does the model start?

Answer: The model started approximately 5.11 billion years ago. Calculating out the total timespan for the 7 nights and 7 days period involves multiplying 14 times the length of one of the Lord's days, which gives 5.11 billion Julian years.

Question #5: When does the model end?

Answer: There is a small offset between now and the ending point I am choosing for the model. The ending point for the model is Adam and Eve's exit from the garden of Eden, which was about 6,000 to 10,000 years ago based on adding up the ages of the genealogies and estimating roughly for generation time. For most of my calculations, I will be discarding the approximately 6,000 to 10,000 year timespan. This discarding is acceptable because that timespan is relatively small compared to the scientific ages, which are mostly in millions and billions of years.

Chapter 9: Building the Model — Justification for the First Unit of the Model being a Night and Justification for the Requirement that No Creation Events Described as being in our Solar System Occurred at Night

This chapter will justify the answer that I gave to question #6 in Chapter Four.

Question #6: Is the first unit a Day or a Night?

Answer: The first unit is a Night. The sequence goes Night, Day, Night, Day, etc. The text of Genesis Chapter 1 verse 5 and other knowledge support this statement.

In Young's Literal Translation of the Bible, the evening leads first in Genesis 1:5.

---

and God calleth to the light 'Day,' and to the darkness He hath called 'Night;' and there is an evening, and there is a morning--day one. (Genesis 1:5, YLT)

So the Night goes first in the sequence.

Although Genesis 1:5 should be sufficient on this point, Jewish tradition also supports a night-to-night timeframe being preferable. In traditional Jewish practice, the time frame's demarcation for a day is sunset-to-sunset, not sunrise-to-sunrise.[24] In other words, according to the Biblical structure, the evening leads. The Night goes first, and thus the sequence is {Night, Day, Night, Day, Night, Day, etc.}.

All the ancient events in our solar system that the Bible described in the Genesis Creation Account occurred during the Days. This is in accordance with the passage in Exodus 20:11 YLT for God (Jehovah) made the heavens and the earth during the first six days. Below is the passage.

---

for six days hath Jehovah made the heavens and the earth, the sea, and all that is in them, and resteth in the seventh day; therefore hath Jehovah blessed the Sabbath-day, and doth sanctify it. (Exodus 20:11, YLT)

Notice that Moses recorded God as having said 'for six days'. Since daytime and nighttime are correctly being treated as being distinct to the Lord, and the text stated the making was done for 'six days', the events are required to occur during the Lord's Daytime. In this book, each of the 13 ancient events occurred during the daytime. This will later play a role in calculating the probability of an event occurring in a distinct time slot, as an event landing on nighttime would contradict the above statement.

Chapter 10: The Model Timeline

My goal was to construct a testable model for the Genesis Creation Account. I wanted a timeline that could be broken down into time windows with which I could test the Genesis Creation Account against the known timings of ancient scientific events.

Both the Gregorian Calendar that is used currently throughout the world as well as the Julian Calendar of Saint Peter's lifetime use 365 days per year for a non-leap year. A leap year is a year that contains 366 days. The frequency of leap years compared to normal 365-day years differs between the two calendar systems. For the Julian Calendar, a leap year is exactly once every fourth year. This leads to the Julian Calendar having on average 365.25 days per year. Due to its simplicity, scientists often use the Julian Calendar years for calculations in the past. This leads to a difference between the two Calendar Systems that is insignificant for my purposes. See Appendix A for a full description and the resulting mathematical effect on the time calculations used in this book. To make the book more readable and for ease of calculation for the reader, I ignore the leap year's contribution to the days per year throughout most of this book. This simplification on its own induces an underestimate of about 3,500,000 years for the total time block in my calculated years. Since I am primarily concerned with the first 3 significant figures, and since the difference is very minor as a percent of the total 5.11 billion years (less than 0.1% error), the error introduced due to this relaxation of rigor by ignoring the leap year and the imprecision of mismatching calendars is tolerable to the timeline.

With 14 units within the model, each unit being about 365,000,000 Julian years long, I could determine the total amount of time for the model as 5,110,000,000 years ago (i.e. 5.11 billion years ago). Since the model's endpoint is about 6,000 to 10,000 years ago, we can ignore that time period as being insignificant to the scale of the calculation. This simplification does not change the model's beginning point; the beginning of Night One was about 5,110,000,000 years ago (i.e. 5.11 billion years ago). The first unit is a Night. Altogether, these statements generate the following table: Table 2.

TABLE 2: The Timetable of Genesis			
Timeline in Genesis 1 & 2	**Nights and Days that Remain until the End of Day 7**	**Units of the Lord's Time that Remain until the End of Day 7**	**Calculated Timeline Based on 2Peter 3:8 (years ago)**
Beginning of Night One	7 nights + 7 days	14	5,110,000,000 (5.11 billion years ago)
Beginning of Day One	6 nights + 7 days	13	4,745,000,000
Beginning of Night Two	6 nights + 6 days	12	4,380,000,000
Beginning of Day Two	5 nights + 6 days	11	4,015,000,000
Beginning of Night Three	5 nights + 5 days	10	3,650,000,000
Beginning of Day Three	4 nights + 5 days	9	3,285,000,000
Beginning of Night Four	4 nights + 4 days	8	2,920,000,000
Beginning of Day Four	3 nights + 4 days	7	2,555,000,000
Beginning of Night Five	3 nights + 3 days	6	2,190,000,000
Beginning of Day Five	2 nights + 3 days	5	1,825,000,000
Beginning of Night Six	2 nights + 2 days	4	1,460,000,000
Beginning of Day Six	1 night + 2 days	3	1,095,000,000
Beginning of Night Seven	1 night + 1 day	2	730,000,000
Beginning of Day Seven	1 day	1	365,000,000
End of Day Seven	None	0	0 (6,000 to 10,000 years ago)
The fourth column is the result of multiplying the third column by 365,000,000 years			

To calculate the beginning point to the ending point for each Night and for each Day, take the sum of the remaining days and nights of the Lord's Time (i.e. take the sum of a cell in column 3 in Table 2) and multiply by the length of one of the Lord's Days (365,000,000 years). An alternative approach would be to subtract a block of 365,000,000 years repeatedly from the total 5,110,000,000 years. Both methods yield the same outcome; both methods calculate the historical timeline for the Genesis Creation Account.

Table 2 was composed using a mathematical interpretation of 2Peter 3:8. The second and third columns describe and provide the math for corresponding the time slot in the first column with the time enumerated in the fourth column. Each time slot's beginning point also marks the ending point of the previous time slot, so the time serves as the boundary point between the two time slots. For example, 730,000,000 marks the beginning point of Night Seven as well as the ending point of Day Six. Likewise, Night Seven spans from 730,000,000 years ago to 365,000,000 years ago with the later time point marking the beginning of Day Seven.

Since we have seen the connection between the beginning of a time slot and its time in years, Table 2 can be simplified by removing the second and third columns and by removing the bottom note. These simplifications give us Table 3.

Table 3: The Simplified Timetable for the Beginning of Genesis

My characterization of time slots by each slot's starting point based on Genesis 1 and 2	**Calculated Timeline Based on 2Peter 3:8 (years ago)**
Beginning of Night One	5,110,000,000
Beginning of Day One	4,745,000,000
Beginning of Night Two	4,380,000,000
Beginning of Day Two	4,015,000,000
Beginning of Night Three	3,650,000,000
Beginning of Day Three	3,285,000,000
Beginning of Night Four	2,920,000,000
Beginning of Day Four	2,555,000,000
Beginning of Night Five	2,190,000,000
Beginning of Day Five	1,825,000,000

Beginning of Night Six	1,460,000,000
Beginning of Day Six	1,095,000,000
Beginning of Night Seven	730,000,000
Beginning of Day Seven	365,000,000
End of Day Seven	0

Table 3 presents the core information that the reader needs to know to begin testing the idea of whether the Genesis Creation Account is consistent with Science or not.

Another way of visualizing the information is as a timeline.

Timeline for Seven Nights and Seven Days in Genesis 1 and 2

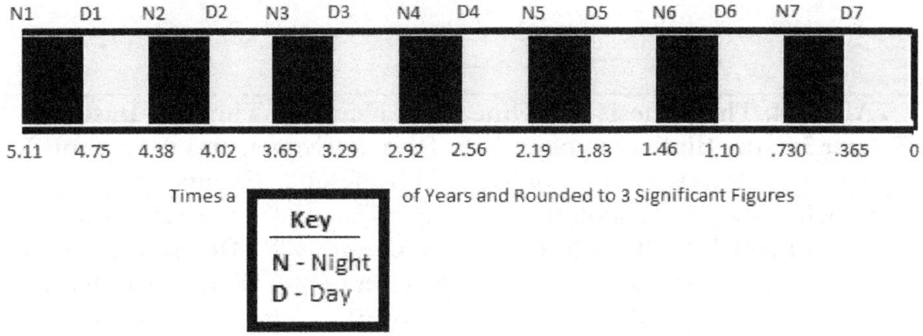

If the Genesis Creation Account is truthful and God-inspired, then we would expect the information regarding the events depicted in the account to fall on the scientifically correct testable time windows. If the Genesis Creation Account is a lie, a fiction invented by clever-but-not-divinely-inspired men thousands of years ago, we would expect failure when attempting to correlate the Genesis Creation Account with modern scientific data. Further, to strongly support the existence of God, the event correlation must be shown to be strong. For all events with a time that can be determined, there must be no strong contradiction between the Genesis Creation Account's timing for an ancient event and the scientific time determination for the same ancient event.

If the Genesis Creation Account is nonsense, a single event proven to occur at a specific time outside of its Bible-predicted time slot would be enough. Note that only a knowledge-based contradiction, not a popularity-based contradiction, would have such an effect.

But if the Genesis Creation Account is shown to be true -- if we cannot find a single strong contradiction, the utter improbability of not having a single provable event against the testimony of the Book of Genesis must be accounted in God's favor and must be acknowledged.

Table 4 is the table format that I will be using to demonstrate the correlation of the Biblical Creation Account descriptions to the scientifically determined events. I will be filling in this table after each event is discussed in Chapters 11 - 28. Table 4 has two more columns than Table 3: one for the Biblical Account Reference Events and one for the Scientific Determination of Time for each Reference Event plus citations.

TABLE 4: The Genesis Timeline, the Calculated Timeline Based on 2Peter 3:8, the Biblical Account's Reference Events, and the Scientific Determination of Time for Reference Events.

Timeline in Genesis 1 and 2	Calculated Timeline Based on 2Peter 3:8 (years ago)	Biblical Account Reference Event	Scientific Determination of Time for Reference Event (plus citations)
Beginning of Night One	5,110,000,000 (5.11 Billion)		
Beginning of Day One	4,745,000,000		
Beginning of Night Two	4,380,000,000		
Beginning of Day Two	4,015,000,000		
Beginning of Night Three	3,650,000,000		
Beginning of Day Three	3,285,000,000		
Beginning of Night Four	2,920,000,000		
Beginning of Day Four	2,555,000,000		
Beginning of	2,190,000,000		

Night Five			
Beginning of Day Five	1,825,000,000		
Beginning of Night Six	1,460,000,000		
Beginning of Day Six	1,095,000,000		
Beginning of Night Seven	730,000,000		
Beginning of Day Seven	365,000,000		
End of Day Seven	0		

Now that we have testable time windows, in the next chapter, we will start filling in Table 4 to begin the comparison of the scientific data to the timeline for the Genesis Creation Account — starting with the Earth.

Chapter 11: DAY ONE — The Creation of the Earth

According to the Bible, the Earth existed on Day One.

In the beginning of God's preparing the heavens and the earth-- the earth hath existed waste and void, and darkness is on the face of the deep, and the Spirit of God fluttering on the face of the waters, and God saith, `Let light be;' and light is. (Genesis 1:1-3, YLT)

As a way of reminder to the reader, Day One began at 4,745,000,000 years ago, according to our model, and ran until 4,380,000,000 years ago, according to the model.

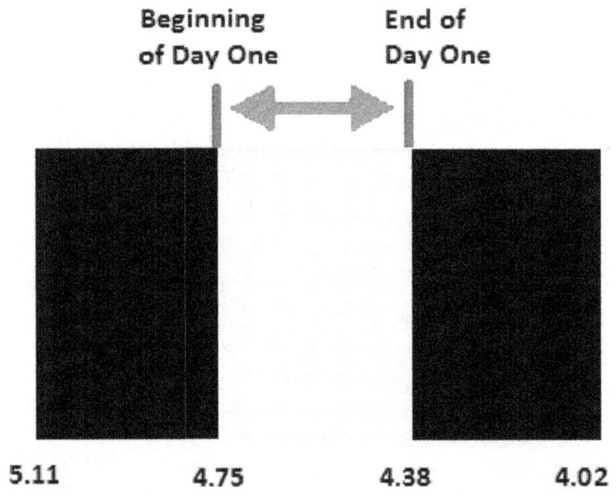

Times are given in Billions of Years and rounded to 3 Significant Figures

So the first observation is that the timing for Earth's creation is required to fall within that window. Another constraint from the first three verses of Genesis is that the Earth's existence must precede the light, which I am identifying as the Sun. So the Earth existed before the Sun, according to the Bible. The Sun's age is known to be 4.57 billion years old.[6] The relative ordering of the Earth and the Sun is the first challenge because a popular notion is that the Sun originated first and that the Earth came later.

The question is "How do we know that the Earth existed before the Sun?" We have some confidence in the Earth having existed before the Sun from radiometric samples taken from the lead on Earth, from the narrative of our Moon's creation, and from physical samples from the Moon.

Provided below is a table of the names of several scientists who have supported the age of the Earth as being 4.65 billion years ago.

TABLE 5: List of Scientists who have made a statement supporting an age of 4.65 billion years for the Earth				
Name of Scientist	**Field or Fields**	**Association**	**Statement supporting 4.65 billion-year-old Earth**	**Ref #**
Niles Eldredge	Biology, Paleontology	Columbia University	"As we now know, Darwin and Lyell were quite right: the earth is 4.65 billion years old, and life goes back as far as at least 3.5 billion years – in the form of tiny bacteria that are the oldest fossils so far discovered."	25
David J. Cuff	Physical Geography	Temple University	"The Earth and other nearby planetary bodies formed approximately 4.65 billion years ago as a result of the clustering of condensed particles early in the history of the solar system."	26
David H. Cornell	Geochemistry, Geology	Gothenburg, Sweden	"Thus as shown by Figure 3, pre-solar system processes are cancelled out, allowing us to concentrate on those which have	27

			operated since the earth formed some 4650 million years ago.”	
Leonard G. Wilson	Medicine, Biology	Head of the Department of the History of Medicine of the University of Minnesota.	(Leonard wrote biographies about the geologist Charles Lyell.) Recipient of 2013 Mary C. Rabbitt History And Philosophy of Geology Award from The Geological Society of America. While accepting the award, Leonard G. Wilson stated: “Today the Earth is calculated to be some 4,650 million years old. Throughout almost all of that time, the Earth, though constantly active, has remained remarkably stable.”	28
Wolfgang Schlager	Geology	Professor Emeritus at The Vrije Universiteit Amsterdam in the Netherlands.	Under the heading ‘2. EARTH’S PHYSICAL LAYERS AND THEIR CYCLES’ is the following: “The masses that condensed into our planet about 4.65 billion years ago soon differentiated into layers of different density under the influence of gravity.”	29
Derek E.G. Briggs	Paleontology	A Professor of Earth & Planetary Sciences at Yale	“We now know, through multiple lines of evidence, that our planet formed 4.65 billion years ago and	30

		University; Curator in Charge of Invertebrate Paleontology, Yale Peabody Museum of Natural History	that conditions suitable for life were present relatively soon afterward (p. 90)."	
Alan Collins	Tectonic Geology	Professor at the University of Adelaide in Australia.	"I have always loved maps and been interested in the world. Mapping the globe over its 4.65 billion year history is a natural extension of this to me."	31
Edward Peter Jacobus van den Heuval	Astronomy	Professor at the University of Amsterdam	From the online Abstract of the section *From Universe to Multiverse* for the book titled *The Amazing Unity of the Universe*: "In the first place there is the fact that the age of the universe, 13.8 billion years, does not differ much from the age of the solar system and Earth of 4.65 billion years, and from the time of about 4 billion years that life on Earth needed to produce observers that can think about the laws of nature and about the evolution of the universe."	32

There are also scientists that think or have thought that 4.7 billion years old is the correct age for the Earth.

TABLE 6: List of Scientists who have made a statement supporting an age of 4.7 billion years for the Earth				
Name of Scientist	Field or Fields	Association	Statement supporting 4.7 billion-year-old Earth	Ref #
Melvin Calvin	Chemistry, Biochemistry	Professor at the University of California, Berkeley. Nobel Prize Laureate in Chemistry.	"The earth is 4700 million years and the Soudan is dated at > 2700 million years."	33
Peter Cattermole	Geology	Former Principal Investigator with NASA's Planetary Geology and Physics Program.	"But so far as the Earth is concerned, we have at least a few firm facts to guide us, and in particular we are reasonably confident about the time scale. The Earth began its separate existence about 4700 million years ago; the universe is older still, and a reasonable estimate leads us to a figure of around 15,000 million years."	34
Stuart Burley	Geology	Professor of Petroleum Geology, Keele University	"We know the age of the earth, with reasonable certainty and accuracy. It's around 4.7 billion	35

			years old, as determined from the decay of radioactive elements"	
Eli K. Moore	Environmental Science	School of Earth and Environment, Rowan University, Glassboro, NJ, United States	"Written in the R open-source language (R Core Team, 2019) using the shiny package (Chang et al., 2020), dragon is an interactive browser-based application that allows users of any disciplinary background to explore, manipulate, visualize, and statistically analyze mineral-chemistry networks at a userselected time range within the 4.7 billion year history of the mineral record of Earth and extraterrestrial sources (i.e., meteorites, asteroids, pre-solar sources)."	36
Geoffrey Boulton	Geology	Professor of Geology, Edinburgh University.	"After the Earth's formation at 4700 Ma ago, solidification of surface rocks and volcanic degassing produced water, carbon dioxide, nitrogen, and hydrogen sulphide on the surface."	37

It is noteworthy that at two significant figures, 4.65 rounds to 4.7. And some scientists sometimes give timings for the age of the Earth slightly older than 4.65 like A. Krishna Sinha and George R. Tilton. Looking at lead samples from Earth in 1973, they decided that, "...The best estimate of the age of the Earth from present lead data is 4660 m.y., an age in remarkable agreement with that proposed for the Moon (4650 m.y.) and newer measurements on chondritic meteorites (4640 m.y.)."[38]

So, where does the Moon come in? The Moon's creation acts as a minimum timing for the Earth's existence. The famous scientist Brent Dalrymple remarked in his book Ancient Earth, Ancient Skies: The Age of Earth and Its Cosmic Surroundings that, "The only viable hypothesis for the formation of the Moon is that it aggregated from the debris formed when a planetoid about the size of Mars or larger collided with a proto-Earth only about half its present size. If this hypothesis is correct, then the minimum age for the Moon is also a minimum age for Earth, and it follows that Earth and the Moon must be at least 4.5 billion years in age."[39] In other words, you cannot have a collision between a Mars-sized object and the proto-Earth if the proto-Earth did not yet exist. Thus, the Moon's formation time acts as a minimum time for the existence of the proto-Earth. I will use Mr. Dalrymple's reasoning, namely that if the Moon is older than the Sun, that will imply that the Earth is also older than the Sun.

So if the Moon's age is a minimum timing for the Earth, it raises the question, "How old is the Moon exactly?" There was disagreement among scientists regarding the answer to this question. For some of the scientists who relied on the soil and breccia gathered during the Apollo missions, their works state 4.60 to 4.7 billion years, which would make the Moon older than the Sun. For those who relied on the work done by Laurence E. Nyquist, then the answer would be a Moon younger than the Sun.

So back to the question, "How old is the Moon?" The work of some government geologists that represented the United States government in a conference with scientists from the USSR in 1974 would give an age for the Moon older than the 4.568 billion-year-old Sun. In the *Early History of the Moon: Implications of U-Th-Pb and Rb-Sr Systematics*, U.S. Geological Survey scientists Mitsunobu Tatsumoto, Paul D. Nunes, and Daniel M. Unruh remarked in the abstract to their paper that, "We hypothesize from these and other physical, chemical, and petrographic results that (1) the Moon formed ~ 4.65 b.y. ago;..."[40] They also remarked that, "As discussed above, all U-Pb and Rb-Sr data obtained from lunar samples are compatible with this primary age of ~ 4.65 b.y".[41] They are not alone in their thinking. Here is a list of some more scientists who gave an age for the Moon that is older than the 4.568 billion-year-old Sun.

TABLE 7: The Names of Scientists, Their Professional Positions, and the Times They gave for at least some Moon-related Samples with the Citation to their work.			
Name of Scientist	**Position of Scientist or government representative held at some point in their careers**	**Time given for at least some Moon-related samples (billions of years ago)**	**Citation of work**
C. E. Hedge B. R. Doe	US Department of the Interior's U.S. Geological Survey	Lunar formation at about 4.65	42
K. Gopalan S. Kaushal C. Lee-Hu G. W. Wetherill	Emeritus Scientist at National Geophysical Research Institute in India. Associated with the University of California at Los Angeles Institute of Geophysics and Planetary Physics Associated with the University of California at Los Angeles Institute of Geophysics and Planetary Physics Director Emeritus of the Department of Terrestrial Magnetism of the Carnegie Institution of Washington, DC.	4.7 from surface material[43]	44

D.A. Papanastassiou	Senior Research Scientist at California Institute of Technology's Jet Propulsion Laboratory, which is associated with NASA	4.72 model age for lunar soil sample 14259	45
Leon T. Silver	Professor of Geology at Caltech	4.60 to 4.63 from Moon dust and breccia	46
John N. Rosholt	U. S. Geological Survey, Denver, Colorado	4.66 from fine dust and breccia, sample 10084 and sample 10061, respectively, based on $^{207}Pb/^{206}Pb$, $^{206}Pb/^{238}U$, $^{207}Pb/^{235}U$, and $^{208}Pb/^{232}Th$ ratios.[47]	48
Paul D. Lowman, Jr.	NASA geophysicist associated with Goddard Space Flight Center in Greenbelt, MD	4.7 origin of Moon	49
William Ian Ridley	Associated with Columbia University and affiliated with NASA's University Program. Later, Research Scientist for the U.S. Department of the Interior.	'The Moon accreted homogeneously at ~ 4.65 AE in a short time period (<100 yr)'	50
James Fred Hays David Walker	Department Chairman of Geological Sciences at Harvard University Higgins Professor Emeritus at Columbia University in Earth and Environmental Sciences Department	4.7 given in statement regarding lunar petrologic history	51

If any of the cited Table 7 scientists were correct regarding the Moon's age, you get an Earth at least that old, and thus older than the Sun. Based on Tables 5 and 6 and the above scientists, the Earth is around 4.65 billion years old, which rounds up to 4.7 billion years when given as two significant figures. This observation makes the Earth's placement on Day One correct as that 4.745 rounds at three significant figures to 4.75, which accommodates a 4.7 timing. That time range is older than the 4.568 billion year age for the Sun, making the Earth's relative positioning before 'the light' correct as well.

So the Earth was the first reference event examined, and its age is supportive of the Biblical timeline. Below in Table 8, I have included a copy of Table 3 filled-in with an entry corresponding to the Earth and its timing.

TABLE 8: The Genesis Timeline, the Calculated Timeline Based on 2Peter 3:8, the Biblical Account's Reference Events, and the Scientific Determination of Time for Reference Events Thru the Earth in Chapter 11.

Timeline in Genesis 1 & 2	Calculated Timeline Based on 2Peter 3:8 (years ago)	Biblical Account Reference Event	Scientific Determination of Time for Reference Event (plus citations)
Beginning of Night One	5,110,000,000 (5.11 Billion)		
Beginning of Day One	4,745,000,000	'the earth'	The earth formed at about or slightly before 4.65 billion years ago.[4]
Beginning of Night Two	4,380,000,000		
Beginning of Day Two	4,015,000,000		
Beginning of Night Three	3,650,000,000		
Beginning of Day Three	3,285,000,000		
Beginning of Night Four	2,920,000,000		
Beginning of Day Four	2,555,000,000		
Beginning of Night Five	2,190,000,000		

Beginning of Day Five	1,825,000,000		
Beginning of Night Six	1,460,000,000		
Beginning of Day Six	1,095,000,000		
Beginning of Night Seven	730,000,000		
Beginning of Day Seven	365,000,000		
End of Day Seven	0		

Chapter 12: DAY ONE — The Deep

Next, I will discuss the presence of 'the Deep' before the formation of the Sun. I am identifying 'the deep' as a large supply of space ice that existed before the birth of the Sun. Although I do not know how long the large supply of space ice pre-existed the Sun, I am content to remark on its presence being consistent with and supportive of the Biblical Creation Account. Since the large supply of space ice may not be a single point of time but rather may have been around for a long time, I will not be including this occurrence in my 13 ancient events total with regards to probability. Here is the relevant passage from Genesis Chapter One that describes 'the deep' before 'the light.'

In the beginning of God's preparing the heavens and the earth-- the earth hath existed waste and void, and darkness is on the face of the deep, and the Spirit of God fluttering on the face of the waters, and God saith, `Let light be;' and light is. (Genesis 1:1-3, YLT)

How do we know that the deep existed and was present before the Sun's birth? We know through a set of inferences drawn that some of Earth's water predates the Sun. *The Ancient Heritage of Water Ice in the Solar System* was a scientific work authored by a team of scientists led by Ilsedore L. Cleeves and representing institutions such as the University of Michigan, Carnegie Institution of Washington, Harvard University, and the University of Exeter.[5] The scientists found that for a high calculated value of deuterium relative to the hydrogen in water in the dense interstellar medium that "...at the very least..." approximately 7% or more of the Earth's ocean water should predate the sun, and that for a lower value, more would: 30% to 50% of the Earth's ocean water.[52] The discovery that some of the Earth's water is older than the Sun resulted in media coverage and articles. Examples include "Much of Earth's Water is Older Than the Sun" on space.com by Senior Editor Mike Wall and the "Earth's Water Is Older Than the Sun" on discovermagazine.com on Carl Engelking's blog. An electronic article titled "The Water in Your Glass Might Be Older Than the Sun" by Nicholas St. Fleur appeared in the Science Times section of the New York Times website.

This electronic article had a comment that a version of the article appeared on page D2 on April 26, 2016, in the New York edition.[53] [54] [55] At the beginning of the scientific work by Cleeves et al, the scientists posed an open question of where the ices came from. The team of scientists considered two scenarios regarding the ices: "...whether they i) originated in the dense interstellar medium (ISM) in the cold molecular cloud core prior to the Sun's formation, or ii) are products of reprocessing within the solar nebula (5–7)".[56] The scientists' research ruled out the scenario that all the ice originated within the solar nebula's protoplanetary disk. Consequently, the scientists decided that at least some of Earth's water must have originated from a parent molecular cloud core that existed before the Sun's formation.

So a parent molecular cloud core from before the Sun's formation existed. The scientists also wrote, "These findings imply that some amount of interstellar ice survived the formation of the solar system and was incorporated into planetesimal bodies."[52] The authors considered the first scenario to imply that "...abundant interstellar ices, including water and presolar organic material, are incorporated into all planet-forming disks."[57] Therefore, all the planet-forming disks were affected. And Earth is one of them.

The parent molecular cloud core was present before the formation of the Sun. This statement means that ice would have been present in space before the Sun's formation. So the existence of 'the deep' and 'the face of the waters' before the Sun's formation are correct.

Starting with Table 9, the table data will be truncated at the mark of two Nights below the event being described.

TABLE 9: The Genesis Timeline, the Calculated Timeline Based on 2Peter 3:8, the Biblical Account's Reference Events, and the Scientific Determination of Time for Reference Events Thru The Deep in Chapter 12.			
Timeline in Genesis 1 & 2	**Calculated Timeline Based on 2Peter 3:8 (years ago)**	**Biblical Account Reference Event**	**Scientific Determination of Time for Reference Event (plus citations)**
Beginning of Night One	5,110,000,000 (5.11 Billion)		
Beginning of Day One	4,745,000,000	'the earth'	The earth formed at about or slightly before 4.65 billion years ago.[4]
		'the deep' and 'the face of the waters'	Space ice supply predating the Sun[5]
Beginning of Night Two	4,380,000,000		
Beginning of Day Two	4,015,000,000		
Beginning of Night Three	3,650,000,000		

Chapter 13: DAY ONE — The Light as the Sun

I will discuss 'the light' from Genesis 1:3-5 as being the Sun. I will be discussing how the Sun's age compares to the time window for Day One. Here is the relevant passage from Genesis Chapter One referring to 'the light' on Day One.

... and God saith, `Let light be;' and light is. And God seeth the light that it is good, and God separateth between the light and the darkness, and God calleth to the light `Day,' and to the darkness He hath called `Night;' and there is an evening, and there is a morning--day one. (Genesis 1:3-5, YLT)

To determine where this event's timing is consistent with the timeline we generated for Genesis using primarily Saint Peter's statement, we should remind ourselves of the timeline constructed for Genesis. The Biblical time window allowing for Day One events to occur spans from the beginning of Day One at 4,745,000,000 Julian years ago until 4,380,000,000 Julian years ago. Here is a diagram to remind us of the timeline's day one time window.

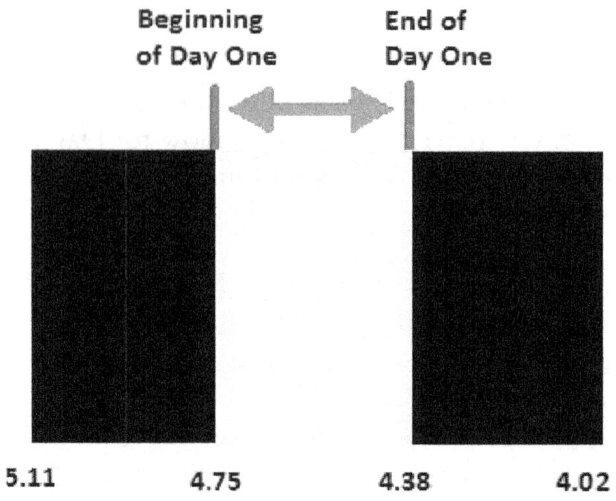

Times are given in Billions of Years and rounded to 3 Significant Figures

For our timeline to remain viable, the Sun's age must fall on or after the beginning of Day One and on or before the end of Day One. An exacting age of the Sun, "the best estimate for the time of the formation of the Solar System", is set at 4,568.2 million years, which is equal to 4.5682 billion years ago.[6] The Sun's age has an uncertainty of up to 0.0002 billion years on the positive side and up to -0.0004 billion years on the negative side.[58] Arizona State University researchers Dr. Audrey Bouvier and Dr. Meenakshi Wadhwa supplied this time for the Sun in their letter to Nature Geoscience.[59] Another estimate of the Solar System's age, and thus the Sun's age, was given by Amelin et al using an uranium-to-uranium radiometric dating for the Allende meteorite, which put the beginning of the Sun's coming together at around 4567.18±0.50 million years ago, which is equal to a timing of 4.56718 ± .00050 billion years ago.[60] A less exacting age of the Sun's timing is given by Dunford on a NASA website to be 4.5 billion years old[61], but let us stick with the more exact timing from Bouvier and Wadhwa.

According to the calculated timetable for Genesis, Day One's time window begins at approximately 4,745,000,000 years ago and runs until approximately 4,380,000,000 years ago. Notice that the Sun's scientific timing at 4,568,200,000 places it securely on Day One. That is, the Sun's age is after the beginning of Day One and before the beginning of Night Two. The Sun's scientific age falls within the calculated interval for Day One in the

Bible. The Bible's timeline for the reference event of the creation of the Sun is consistent with modern science.

An inquisitive reader who is unfamiliar with astronomy and radiometric dating might inquire as to how the scientists were able to determine the age of the Sun. Famed scientist Brent Dalrymple indicated that 'chemical considerations' showed that a number of small inclusions 'of unusual high-temperature minerals' on the Allende meteorite must have condensed directly from the Solar Nebula.[62] An inclusion is constituted of "Fragments of older material (e.g. Minerals, rock) that have been enclosed within another rock."[63] The inclusions on the Allende meteorite are examples of calcium-aluminum-rich inclusions, a term that is abbreviated in scientific literature in the plural as 'CAIs' and in the singular as 'a CAI'. Takashi Yoshizakia and a group of scientists from Japan and South Korea wrote the following about calcium-aluminum-rich inclusions. CAIs are, "...predicted by thermodynamic models to be among the first solids to condense from a cooling gas of the solar composition (e.g., Grossman, 1972). They are considered to have formed during high-temperature processes including condensation, evaporation, aggregation and melting that occurred near the proto-Sun (e.g., McKeegan et al., 2000; Sossi et al., 2017). Shortly after their formation, CAIs were transported to the asteroid- or even comet-forming region (e.g., Zolensky et al., 2006) via X-winds (e.g., Shu et al., 1996), outwards turbulent flow (e.g., Ciesla, 2007), or disk winds (Van Kooten et al., 2016)."[64]

From what I've read, it appears to me that Bouvier and Wadhwa used a dating technique called lead-lead dating on a calcium-aluminum-rich inclusion to determine its age, while Amelin et al used uranium-uranium dating on a calcium-aluminum-rich inclusion to determine its age.

TABLE 10: The Genesis Timeline, the Calculated Timeline Based on 2Peter 3:8, the Biblical Account's Reference Events, and the Scientific Determination of Time for Reference Events Thru The Light as the Sun in Chapter 13.			
Timeline in Genesis 1 & 2	**Calculated Timeline Based on 2Peter 3:8 (years ago)**	**Biblical Account Reference Event**	**Scientific Determination of Time for Reference Event (plus citations)**
Beginning of Night One	5,110,000,000		
Beginning of Day One	4,745,000,000	'the earth'	The earth formed at about or slightly before 4.65 billion years ago.[4]
		'the deep' and 'the face of the waters'	Space ice supply predating the Sun[5]
		'the light'	The Sun forms at around 4.57 billion years ago.[6]
Beginning of Night Two	4,380,000,000		
Beginning of Day Two	4,015,000,000		
Beginning of Night Three	3,650,000,000		

Chapter 14: DAY TWO — The Sky

The chief event of Day Two is the formation of the sky. The atmosphere is the sky. Here is the text from Genesis regarding Day Two in the New Heart English Bible:

God said, "Let there be an expanse in the middle of the waters, and let it divide the waters from the waters." God made the expanse, and divided the waters which were under the expanse from the waters which were above the expanse; and it was so. God called the expanse sky. There was evening and there was morning, a second day. (Genesis 1: 6-8, New Heart English Bible)

The text implies that the expanse was brought into existence and states that the expanse was called "sky" in the New Heart English Bible translation of the Bible. Some other versions of the Bible use the word "Heaven" or "Heavens" instead of the name sky, such as Darby and Young's Literal Translation, respectively. Suppose that sky is a correct reading. The text can then be turned into an assertion about the atmosphere because the atmosphere is the 'sky.' The assertion is that the atmosphere must have come into existence during Day Two. That means that the atmosphere's coming into being would need to fall between 4,015,000,000 years ago and 3,650,000,000 years ago.

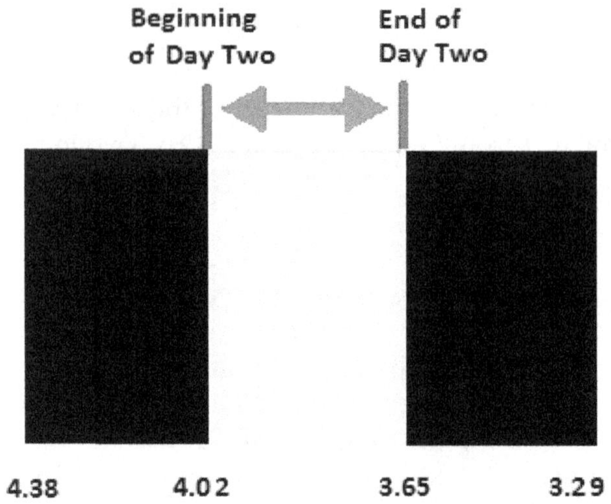

Times are given in Billions of Years and rounded to 3 Significant Figures

Now we need to find out the scientific timing for the atmosphere. It is worth noting that the atmosphere at that time had an insignificant amount of oxygen (O_2) so conditions were anaerobic.[65] Using the Sulfur isotope signatures of rocks from the Nuvvuagittuq Greenstone Belt in Canada, Thomassot et al. indicated that [an] "…unusual atmospheric fingerprint has been present since the end of the Hadean, the earliest eon of Earth history."[7] The Hadean Eon's end is defined to be at 4000 Ma by the International Commission on Stratigraphy, which corresponds to four billion years ago chronologically.[8] The critical point is that these Sulfur signatures are chemical indicators of a chemical reaction that is believed to have occurred in the atmosphere of that time period. Thomassot et al.'s paper was published in the Proceedings of the National Academy of Sciences of the United States of America, one of the most reputable scientific journals within the United States.

When taken as a point, 4.0 billion is less than 4,015,000,000 Julian years. However, one might say that they are the same for two significant figures. 4.0 billion years is greater than 3,650,000,000 years ago. Cliff D. Ollier -- an honored Australian geologist and professor emeritus – wrote in an encyclopedia very clearly that: "…the early earth had no atmosphere or hydrosphere."[9] For context and the reasoning, here is more of the passage under the heading 'THE ORIGIN OF THE ATMOSPHERE': "However the earth might have originated, the early earth had no atmosphere or hydrosphere. If it originated at high temperatures, gases (which have an enhanced escape velocity at high temperature) would be lost to space. If it started as a cold body, by accretion of cold rock fragments, it would have no atmosphere until volatiles escape from the rock (by early volcanic outgassing), and no oceans until water in the atmosphere condensed."[9] In agreement with the spirit of Ollier's statement, in my hunt for reliable evidence, I found no reliable evidence for a time earlier than 4.0 billion years for the atmosphere. Thus, the atmosphere appears to have begun at about 4.0 billion years. Its placement falls at the beginning of Day Two, which is in accord with the Genesis Creation account's timeline.

TABLE 11: The Genesis Timeline, the Calculated Timeline Based on 2Peter 3:8, the Biblical Account's Reference Events, and the Scientific Determination of Time for Reference Events Thru The Sky in Chapter 14.			
Timeline in Genesis 1 & 2	**Calculated Timeline Based on 2Peter 3:8 (years ago)**	**Biblical Account Reference Event**	**Scientific Determination of Time for Reference Event (plus citations)**
Beginning of Night One	5,110,000,000		
Beginning of Day One	4,745,000,000	'the earth'	The earth formed at about or slightly before 4.65 billion years ago.[4]
		'the deep' and 'the face of the waters'	Space ice supply predating the Sun[5]
		'the light'	The Sun forms at around 4.57 billion years ago.[6]
Beginning of Night Two	4,380,000,000		
Beginning of Day Two	4,015,000,000	'the sky'	Evidence exists for the atmosphere since the end of the Hadean Eon[7] , which is about 4.0 billion years ago[8] . There is no evidence for an 'early' atmosphere.[9]
Beginning of Night Three	3,650,000,000		
Beginning of Day Three	3,285,000,000		
Beginning of Night Four	2,920,000,000		

Chapter 15: DAY TWO — The Waters under the Expanse as Sustained Liquid Water on Earth

This chapter includes the earliest reliable evidence of liquid water on Earth in accordance with Day Two's description in Genesis. The text of Genesis regarding Day Two in the New Heart English Bible is below:

God said, "Let there be an expanse in the middle of the waters, and let it divide the waters from the waters." God made the expanse, and divided the waters which were under the expanse from the waters which were above the expanse; and it was so. God called the expanse sky. There was evening and there was morning, a second day.

(Genesis 1:6-8, New Heart English Bible)

When the text said, "...and divided the waters which were under the expanse from the waters which were above the expanse; and it was so", the waters or fluids under the sky seem to be a reference to water in the liquid phase. In contrast, the waters or fluids above the expanse could be a reference to gases. If this is so, it is essential to know the answer to the following question, "When was the first sustained presence of liquid water on the face of the Earth?"

The earliest reliable evidence of a sustained presence of liquid water on Earth, which I encountered, was evidence of liquid water-produced sedimentary structures in Greenland. Allen P. Nutman, Stephen J. Mojzsis, and Clark R. Friend recognize and interpret the rocks, the Banded Iron Formations (BIFs), as proof for a hydrosphere and thus a sustained liquid water presence and give a minimum age of approximately 3.85 billion years for these sedimentary structures in Greenland. From their work titled Recognition of ≥3850 Ma water-lain sediments in West Greenland and their significance for the early Archaean Earth,[66] which was published in the scientific journal *Geochimica et Cosmochimica*,[67] Nutman, Mojzsis, and Friend wrote that: "These rocks provide proof that by ~3850 Ma (1) there was a hydrosphere, supporting the chemical sedimentation of BIF, and that not all water was stored in hydrous minerals, and (2) that conditions satisfying the stability of liquid water imply surface temperatures were similar to present."[10]

Encyclopaedia Britannica mirrors my inability to find earlier definitive evidence for the sustained presence of water in liquid phase before 4.0 billion years ago. According to the Encyclopedia Britannica, "The oldest rocks known are 3.9 billion to 4.0 billion years old, and these rocks, though altered by post-depositional processes, show signs of having been deposited in an environment containing water. There is no direct evidence for water for the period between 4.6 billion and 3.9–4.0 billion years ago."[11]

As a reminder, here is the time window for Day Two that we are assessing:

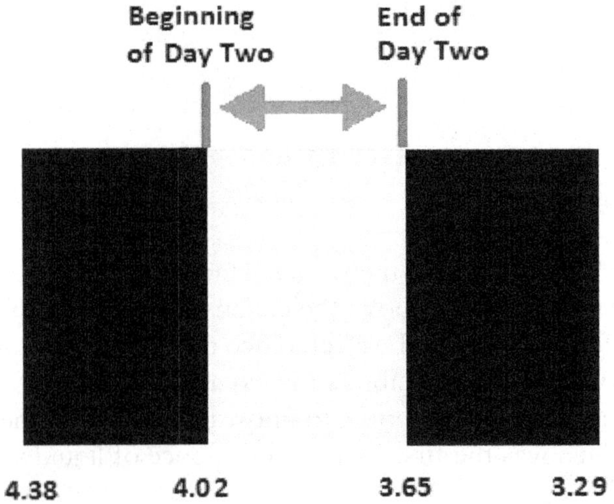

Times are given in Billions of Years and rounded to 3 Significant Figures

The time of 3.85 billion years ago, but not older than 3.9 to 4.0 billion years-old as the Britannica write-up indicated, would limit this almost exclusively to the region at the beginning of Day Two[68]. Consequently, this is a match.

TABLE 12: The Genesis Timeline, the Calculated Timeline Based on 2Peter 3:8, the Biblical Account's Reference Events, and the Scientific Determination of Time for Reference Events Thru the Sustained Liquid Water on Earth in Chapter 15.			
Timeline in Genesis 1 & 2	**Calculated Timeline Based on 2Peter 3:8 (years ago)**	**Biblical Account Reference Event**	**Scientific Determination of Time for Reference Event (plus citations)**
Beginning of Night One	5,110,000,000		
Beginning of Day One	4,745,000,000	'the earth'	The earth formed at about or slightly before 4.65 billion years ago.[4]
		'the deep' and 'the face of the waters'	Space ice supply predating the Sun[5]
		'the light'	The Sun forms at around 4.57 billion years ago.[6]
Beginning of Night Two	4,380,000,000		
Beginning of Day Two	4,015,000,000	'the sky'	Evidence exists for the atmosphere since the end of the Hadean Eon[7], which is about 4.0 billion years ago[8]. There is no evidence for an 'early' atmosphere.[9]
		'God…divided the waters which were under the expanse from the waters which were above the expanse'	The hydrosphere existed by approximately 3.85 billion years ago[10], and there is no direct evidence for a liquid water-based hydrosphere timing before 4.0 billion years ago.[11]

Beginning of Night Three	3,650,000,000		
Beginning of Day Three	3,285,000,000		
Beginning of Night Four	2,920,000,000		

Chapter 16: DAY THREE — The Dry Land that can be Seen

This chapter will cover the first event detailed in the Day Three passage: the dry land that can be seen appearing. Below are two verses detailing this event from Young's Literal Translation of the Bible.

And God saith, `Let the waters under the heavens be collected unto one place, and let the dry land be seen:' and it is so. And God calleth to the dry land `Earth,' and to the collection of the waters He hath called `Seas;' and God seeth that it is good. (Genesis 1:9-10, YLT)

Day Three began at about 3.285 billion years ago and went to about 2.92 billion years ago.

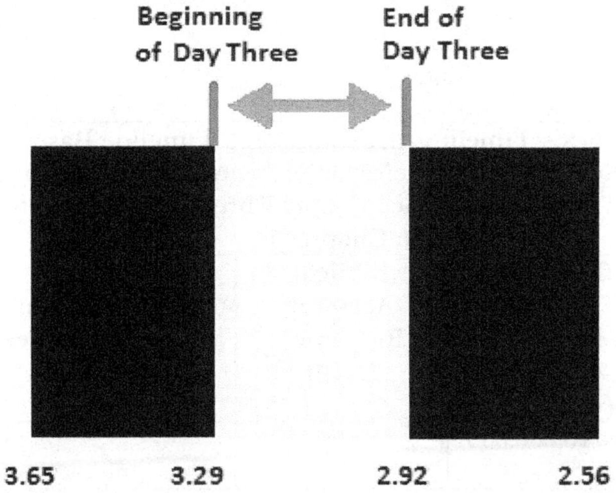

Times are given in Billions of Years and rounded to 3 Significant Figures

The most rigorous answer I could discern from the scientific research is that a landmass appeared above the waterline at about 3.1 billion years ago. The identity of the dry landmass that appeared first is hard to discern, but the possible candidates all had Day Three timings at around 3.1 billion years ago. Before we proceed, two terms should be defined for the reader: paleosol and craton. A paleosol is a fossil soil preserved within a sequence of geological deposits, indicative of past conditions.[69] A craton is the stable interior portion of a continent characteristically composed of ancient crystalline basement rock.[70] With that said, next is the evidence for the approximately 3.1 billion year date for a landmass being above the waterline.

The Singhbhum craton, of which the Keonjhar Paleosol is now a part of, was considered to have achieved being above the waterline at 3.1 billion years ago by some scientists[71]. The Keonjhar Paleosol is located in eastern India, and the Keonjhar Paleosol's age is somewhere between 3.29 billion years old and 3.02 billion years old[72]. Another early instance of rocks that achieved being above the waterline is the Dominion Group of rocks, which preceded the Kaapvaal Craton in southern Africa. The Dominion Group of rocks is about 3.1 billion years old[73] and was clearly above the waterline at 3.1 billion years ago[74 75].

So the Dry Land was above the waterline at about 3.1 billion years ago.[76]

The beginning and ending boundaries for Day Three are 3.285 billion years ago and 2.92 billion years ago, and 3.1 billion years ago is between those two numbers.

TABLE 13: The Genesis Timeline, the Calculated Timeline Based on 2Peter 3:8, the Biblical Account's Reference Events, and the Scientific Determination of Time for Reference Events Thru the Dry Land that can be Seen in Chapter 16.

Timeline in Genesis 1 & 2	Calculated Timeline Based on 2Peter 3:8 (years ago)	Biblical Account Reference Event	Scientific Determination of Time for Reference Event (plus citations)
Beginning of Night One	5,110,000,000		
Beginning of Day One	4,745,000,000	'the earth'	The earth formed at about or slightly before 4.65 billion years ago.[4]
		'the deep' and	Space ice supply

		'the face of the waters'	predating the Sun[5]
		'the light'	The Sun forms at around 4.57 billion years ago.[6]
Beginning of Night Two	4,380,000,000		
Beginning of Day Two	4,015,000,000	'the sky'	Evidence exists for the atmosphere since the end of the Hadean Eon[7], which is about 4.0 billion years ago[8] . There is no evidence for an 'early' atmosphere.[9]
		'God…divided the waters which were under the expanse from the waters which were above the expanse'	The hydrosphere existed by approximately 3.85 billion years ago[10] , and there is no direct evidence for a liquid water-based hydrosphere timing before 4.0 billion years ago.[11]
Beginning of Night Three	3,650,000,000		
Beginning of Day Three	3,285,000,000	'let the dry land be seen'	First landmass to appear above the waterline occurred at about 3.1 billion years ago.[12]
Beginning of Night Four	2,920,000,000		
Beginning of Day Four	2,555,000,000		
Beginning of Night Five	2,190,000,000		

Chapter 17: DAY THREE — The Green as Green Bacteria

The next event I will examine is what the Catholic Public Domain Version of the Bible calls 'green plants' during Day Three's description.

And he said, "Let the land spring forth green plants, both those producing seed, and fruit-bearing trees, producing fruit according to their kind, whose seed is within itself, over all the earth." And so it became. And the land brought forth green plants, both those producing seed, according to their kind, and trees producing fruit, with each having its own way of sowing, according to its species. And God saw that it was good. (Genesis 1:11-12, CPDV)

As a way of reminder, Day Three began about 3,285,000,000 years ago and went until 2,920,000,000 years ago.

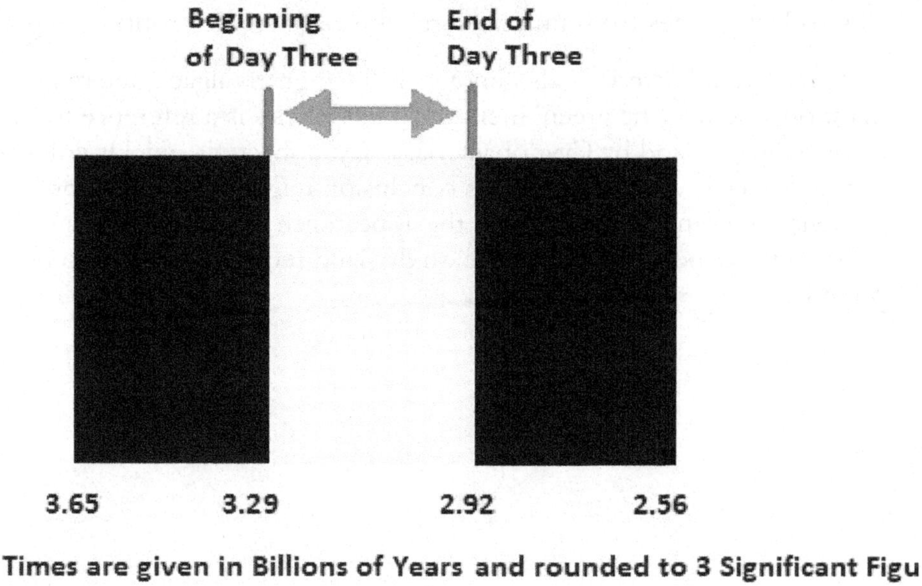

Regarding the Genesis passage, some other Bible translations use words like 'vegetation' or 'herb' in their renderings instead of plants. The root word in Hebrew, deshe', means 'green'[77]. Strong's Definitions for deshe' has 'green' in the list[78]. I will use just 'green' instead of 'green plants', having stopped my parsing at the first word.

Is there any scientific reference event that would match this event – "Let the land spring forth green"? The colonization of land by cyanobacteria might qualify. Cyanobacteria contain chlorophyll and are green.[79] Cyanobacteria is thought to have existed at least 3.48 billion years ago, because we have microbially induce sedimentary structures of microbial mats from the bottom or near the bottom of a body of water from the Dresser Formation in Australia interpreted as dated to that time.[80]

Cyanobacteria is not the only candidate. Chloroflexia, sometimes also called 'Chloroflexi' are the 'green non-sulfur bacteria.' The prefix Chlor- means 'green'[81], while 'flex' means 'bend'. So a green bending. The term coined for land-colonizing bacteria is Terrabacteria. Chloroflexia belongs to this group, and Chloroflexia has a Day Three timing. Chloroflexia is estimated at 3.069 billion years ago with a confidence interval mostly on Day Three as well: confidence interval from 2.922 billion years to 3.184 billion years.[13] This time for the pairwise divergence time for Cyanobacteria and Chloroflexia comes from timetree.org with eight studies contributing to it.[82]

Since Cyanobacteria is also known as 'blue-green algae', one might question whether 'the green' mentioned in Genesis is a reference to the colonization of land by Cyanobacteria. If Cyanobacteria quickly colonized dry land once it was available, this conclusion might follow. The pre-existence of cyanobacteria before the appearance of dry land does not appear to be an issue, because for it to be on dry land requires dry land to be present.

TABLE 14: The Genesis Timeline, the Calculated Timeline Based on 2Peter 3:8, the Biblical Account's Reference Events, and the Scientific Determination of Time for Reference Events Thru The Green in Chapter 17.			
Timeline in Genesis 1 & 2	**Calculated Timeline Based on 2Peter 3:8 (years ago)**	**Biblical Account Reference Event**	**Scientific Determination of Time for Reference Event (plus citations)**
Beginning of Night One	5,110,000,000		
Beginning of Day One	4,745,000,000	'the earth'	The earth formed at about or slightly before 4.65 billion years ago.[4]
		'the deep' and 'the face of the waters'	Space ice supply predating the Sun[5]
		'the light'	The Sun forms at around 4.57 billion years ago.[6]
Beginning of Night Two	4,380,000,000		
Beginning of Day Two	4,015,000,000	'the sky'	Evidence exists for the atmosphere since the end of the Hadean Eon[7], which is about 4.0 billion years ago[8]. There is no evidence for an 'early' atmosphere.[9]
		'God…divided the waters which were under the expanse from the waters which were above the expanse'	The hydrosphere existed by approximately 3.85 billion years ago[10], and there is no direct evidence for a liquid water-based hydrosphere timing before 4.0 billion years ago.[11]

Beginning of Night Three	3,650,000,000		
Beginning of Day Three	3,285,000,000	'let the dry land be seen'	First landmass to appear above the waterline occurred at about 3.1 billion years ago[12]
		'green'	Divergence within land-colonizing bacteria between Cyanobacteria 'blue-green algae' and Chloroflexia estimated at 3.069 billion years ago (CI 2.922 to 3.184 billion years)[13]
Beginning of Night Four	2,920,000,000		
Beginning of Day Four	2,555,000,000		
Beginning of Night Five	2,190,000,000		

Chapter 18: DAY THREE — The Herb Sowing Seed as Firmicutes with Endospores

The next scientific reference event I will discuss corresponds to the earliest 'herb sowing seed'. Below is the Biblical passage.

And God saith, 'Let the earth yield tender grass, herb sowing seed, fruit-tree (whose seed is in itself) making fruit after its kind, on the earth:' and it is so. And the earth bringeth forth tender grass, herb sowing seed after its kind, and tree making fruit (whose seed is in itself) after its kind; and God seeth that it is good; and there is an evening, and there is a morning -- day third. (Genesis 1: 11-13, YLT)

As mentioned in the previous chapter, Day Three began about 3,285,000,000 years ago and went until 2,920,000,000 years ago.

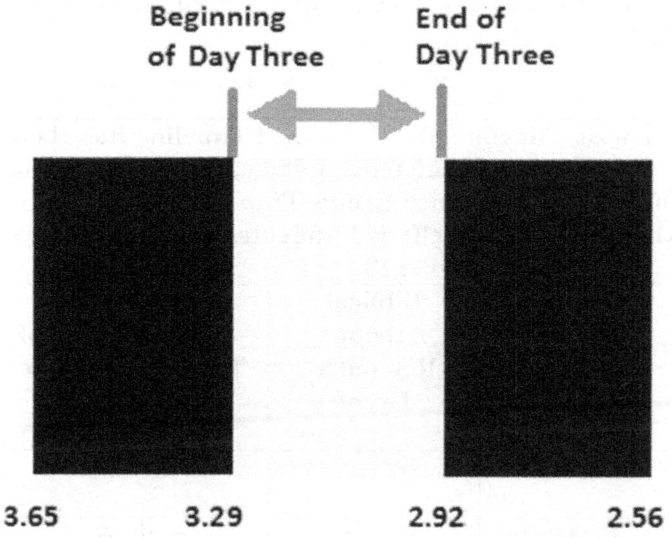

Times are given in Billions of Years and rounded to 3 Significant Figures

Spores can act like seeds: "The term spore derives from the ancient Greek word…spora, meaning "seed, sowing.""[83] Is there anything that colonized land with a Day Three timing that produced spores? Yes, bacteria from the phylum Firmicutes had members that produced endospores[84].

Day Three spans from 3.285 billion years ago to 2.92 billion years ago. Did Firmicutes colonize land during this time? Fabia Battistuzzi, now an associate professor of biology at Oakland University, Andreia Feijao, and Stephen Blair Hedges, director of the Center for Biodiversity in Philadelphia, referred to Firmicutes as belonging to the Terrabacteria – bacteria that colonized land relatively quickly. Timetree.org is a collection of knowledge / a database on the timings of the divergence of organisms and is directed by Sudhir Kumar and Stephen Blair Hedges – both of whom are associated with Temple University. If a website user types 'Firmicutes' and 'Chloroflexia' for the divergence between a pair of organism groupings on the Timetree.org website, 9 studies appear. If one accepts the median time, the timing would be 3.054 billion years ago from Battistuzzi et al's 2004 work titled *A genomic timescale of prokaryote evolution: insights into the origin of methanogenesis, phototrophy, and the colonization of land.*[85] According to *Encyclopaedia Britannica*, a prokaryote is "…any organism that lacks a distinct nucleus and other organelles due to the absence of internal membranes. Bacteria are among the best-known prokaryotic organisms."[86]

TABLE 15: The Genesis Timeline, the Calculated Timeline Based on 2Peter 3:8, the Biblical Account's Reference Events, and the Scientific Determination of Time for Reference Events Thru the Herb sowing seed, fruit-tree (whose seed is in itself) as Firmicutes with Endospores in Chapter 18.

Timeline in Genesis 1 & 2	Calculated Timeline Based on 2Peter 3:8 (years ago)	Biblical Account Reference Event	Scientific Determination of Time for Reference Event (plus citations)
Beginning of Night One	5,110,000,000		
Beginning of Day One	4,745,000,000	'the earth'	The earth formed at about or slightly before 4.65 billion

			years ago.[4]
		'the deep' and 'the face of the waters'	Space ice supply predating the Sun[5]
		'the light'	The Sun forms at around 4.57 billion years ago.[6]
Beginning of Night Two	4,380,000,000		
Beginning of Day Two	4,015,000,000	'the sky'	Evidence exists for the atmosphere since the end of the Hadean Eon[7], which is about 4.0 billion years ago[8]. There is no evidence for an 'early' atmosphere.[9]
		'God…divided the waters which were under the expanse from the waters which were above the expanse'	The hydrosphere existed by approximately 3.85 billion years ago[10], and there is no direct evidence for a liquid water-based hydrosphere timing before 4.0 billion years ago.[11]
Beginning of Night Three	3,650,000,000		
Beginning of Day Three	3,285,000,000	'let the dry land be seen'	First landmass to appear above the waterline occurred at about 3.1 billion years ago[12]
		'green'	Divergence within land-colonizing bacteria between Cyanobacteria 'blue-green algae' and Chloroflexia estimated at 3.069 billion years ago (CI

			2.922 to 3.184 billion years)[13]
		'herb sowing seed'	Estimated median timing for divergence of endospore-forming bacteria phylum Firmicutes is about 3.054 billion years ago.[14]
Beginning of Night Four	2,920,000,000		
Beginning of Day Four	2,555,000,000		
Beginning of Night Five	2,190,000,000		

Chapter 19: DAY THREE — The Fruit-Tree (whose seed is in itself) as Delta-Proteobacteria which has fruiting bodies

I will discuss the reference to 'fruit-tree (whose seed is in itself) making fruit after its kind, on the earth' as a reference to the soil-based bacteria from the delta class of the phylum Proteobacteria. Here is the Biblical passage section from Day Three in Genesis.

And God saith, 'Let the earth yield tender grass, herb sowing seed, fruit-tree (whose seed is in itself) making fruit after its kind, on the earth:' and it is so. And the earth bringeth forth tender grass, herb sowing seed after its kind, and tree making fruit (whose seed is in itself) after its kind; and God seeth that it is good; and there is an evening, and there is a morning -- day third. (Genesis 1: 11-13, YLT)

A curious person or a critic might object with "Hey, what about the tree part?" Both 'tree' and 'stalk' appear as translations in the King James Version from the Hebrew root form 'ets, which is being rendered as 'tree' in the cited passage, but I believe could also be translated as 'stalk'. Both translations for 'ets' appear in the list of usage for 'ets. [87] 'Stalk' is also listed in the outline of Biblical Usage[88] as well as listed under Strong's Definitions in Supplementary Figure 6 in Appendix B.

So supposing that it is acceptable for the phrase to refer to fruit-stalks as opposed to fruit-trees raises the question, "Is there something that has a stalk and exhibits fruiting behavior?" Yes, one species of the delta-proteobacteria is *Stigmatella Aurantiaca. Stigmatella Aurantiaca*'s fruiting body 'consists of a thick stalk supporting a number of individual cysts'[89]. I recommend that the reader goes to Google Images on a computer and types 'Stigmatella Aurantiaca' in the search bar (for copyright purposes, I can't include the images here).

As mentioned in the previous chapter, Day Three began about 3,285,000,000 years ago and went until 2,920,000,000 years ago.

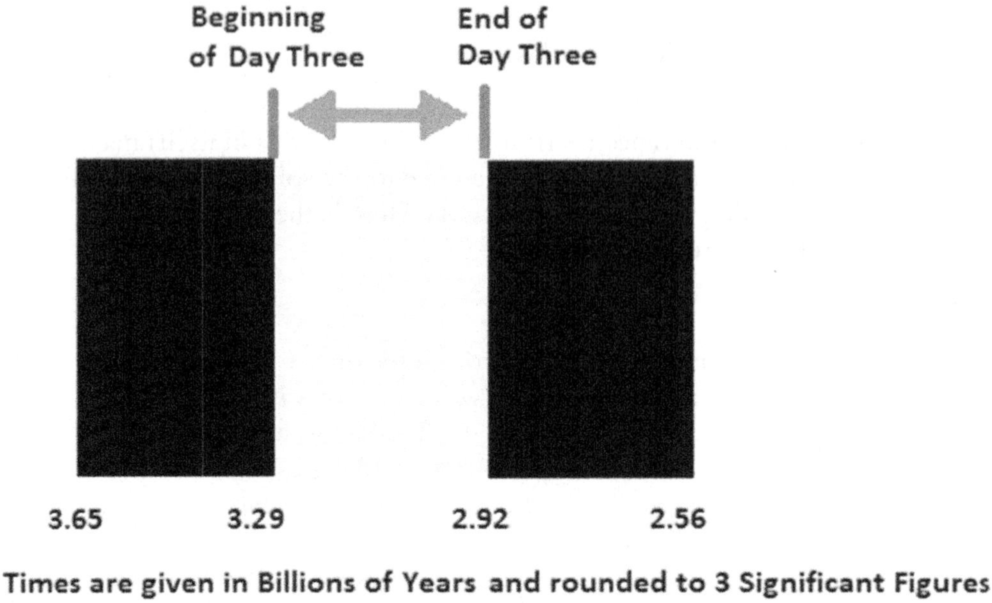

Other Day Three events before "tree making fruit (whose seed is in itself) after its kind", restrict the possible valid timespan from the divergence of Firmicutes to the end of Day Three. This is a time range of about 3.05 billion years ago until about 2.92 billion years ago.

I was unable to determine the time placement of *Stigmatella Aurantiaca*. The research I examined did not all agree on the relative order of which class of proteobacteria came first, which would significantly affect the time placement. Also, when looking at timetree.org and entering Stigmatella Aurantiaca into the 'taxon name', a time of 0.000000000 MYA and a confidence interval of n/a for the Stigmatella genus occurs[90]. It seems to me that the scientists who authored the timetree.org database are not comfortable placing its time. With that said, it seems possible that *Stigmatella Aurantiaca* is being referred to here, but it does not have a reliable time determination.

Chapter 20: DAY FOUR — Addressing and Resolving the Perceived Day Four Contradictions: the Sun, the Moon, and Stars

In the Day Four passage, the following lines are typical for an English translation such as the American Standard Version of the Bible (the ASV):

And God made the two great lights; the greater light to rule the day, and the lesser light to rule the night: he made the stars also. And God set them in the firmament of heaven to give light upon the earth, and to rule over the day and over the night, and to divide the light from the darkness: and God saw that it was good. And there was evening and there was morning, a fourth day. (Genesis 1: 16-19, ASV)

The Sun is a star, and I've already identified 'the Light' on Day 1 as being the Sun. So when the passage appears to state a Day 4 making of the Sun, a making of the Moon, and the making of the stars, these would – if true – imply several contradictions between the passage and what is scientifically known for timing. If left uncorrected, this passage would appear to mean that the Sun was created twice or that my Day 1 placement was incorrect. Let's focus on the critical line: 'And God made the two great lights; the greater light to rule the day, and the lesser light to rule the night: he made the stars also'. Likewise, the Moon pre-existed Day 2 and Day 3 events yet this passage would appear to put it later. There are stars that are older than the Sun[91], yet if they were made on Day 4 and the Sun on Day 1, this order would also appear to lead to a contradiction.

However, all these perceived contradictions can be resolved once the reader knows that the Hebrew word ʿāśâ – which is being rendered as 'made' in the above passage in its Qal verbal stem formation -- can also be rendered as 'appointed'[92]. Rendering ʿāśâ as 'appointed' in the passage resolves the aforementioned perceived contradictions. Consider the below passage with this alternate rendering:

And God appointed the two great lights; the greater light to rule the day, and the lesser light to rule the night: he appointed the stars also. And God set them in the firmament of heaven to give light upon the earth, and to rule over the day and over the night, and to divide the light from the darkness: and God saw that it was good. And there was evening and there was morning, a fourth day. (Genesis 1: 16-19, my modifications to the ASV)

This rendering and its reading avoid the contradictions of a double-creation of the Sun, as well as removes the possibility of a contradiction with the timing of the creation of the stars.

Chapter 21: DAY FOUR — For Signs, for Seasons, for Days and Years

Timekeeping with respect to the stars and constellations and seasonality were instituted on Day Four. Below are the first two verses about Day Four in Genesis using Young's Literal Translation of the Bible:

And God saith, `Let luminaries be in the expanse of the heavens, to make a separation between the day and the night, then they have been for signs, and for seasons, and for days and years, and they have been for luminaries in the expanse of the heavens to give light upon the earth:' and it is so. (Genesis 1:14-15, YLT)

The word 'sidereal' means "of or with respect to the distant stars (i.e. the constellations or fixed stars, not the sun or planets)." So sidereal timekeeping is timekeeping with respect to the distant stars – the constellations or fixed stars.[93] The placement of the constellations or the fixed stars as fixed points is used. In other words, one being able to use 'signs' in the night sky in order to keep time. So when did sidereal timekeeping become possible from earth?

I have not researched this topic exhaustively, but sidereal time-keeping would have been doable since at least 2.45 billion years ago (although Man had not yet been created). By examining patterns in the Weeli Wolli Formation (a banded iron formation in western Australia) and the influence of the tides on the formation, G.E. Williams was able to assign sidereal time for both months per year and for days per year.[94] Williams provides values for both months per year and for days per year with the 'Parameter/Age' listed as 2450 Ma, which indicates that these values correspond to 2.45 billion years ago. This implies that 'for signs…for days and years' has at least a Day Four timing.

I now will address the reference to the seasons. The first seasons according the Biblical Account fall on Day Four, "…then they have been for signs, and for seasons…" Day Four goes from 2,555,000,000 years ago to 2,190,000,000 years ago according to the timetable calculated earlier.

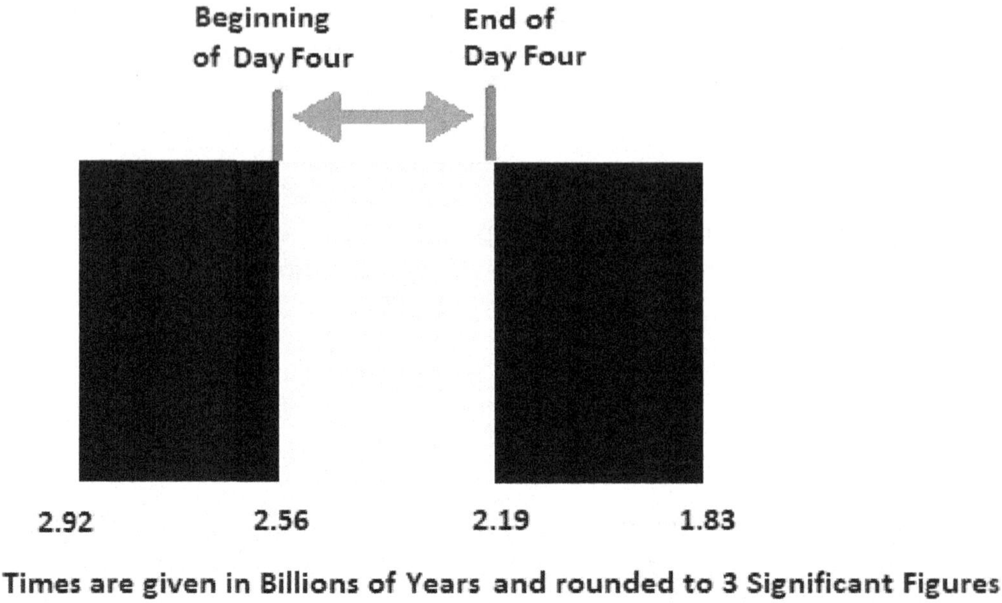

Reliable evidence for strong seasonality is demonstrated from the Huronian glaciation, which occurred between 2.4 billion years ago and 2.3 billion years ago. Both points in time are clearly within the time window for Day Four. The process of being covered by ice sheets or glaciers is termed glaciation. Williams et al wrote the following: "Another persistent feature of Proterozoic glacial successions is the evidence from a variety of facies for a strongly seasonal glacial climate, in particular for large seasonal changes of temperature." For geological evidence for strong seasonality, Williams et al pointed to cryogenic wedges, thaw-modification structures, and Palaeoproterozoic varves. A varve is a pair of thin layers of clay and silt of contrasting color and texture that represent the deposit of a single year (summer and winter) in a lake. Such layers can be used to determine the chronology of glacial sediments.[95]

Williams et al stated that "The oldest known wedge structures interpreted as cryogenic occur in bedded sandstone and siltstone of the lower Huronian

(~2.4 Ga) glaciogenic Ramsay Lake Formation in Ontario, Canada (Young and Long, 1976)."[96] Dr. Williams is cautious when drawing conclusions, and some of the words and his tone reflect that. However, an abundance of caution has not stopped the scientific community from relying on his excellent work, nor will it stop me from citing him as an authority.

TABLE 17: The Genesis Timeline, the Calculated Timeline Based on 2Peter 3:8, the Biblical Account's Reference Events, and the Scientific Determination of Time for Reference Events Thru the Signs, Seasons, Days and Years in Chapter 21.

Timeline in Genesis 1 & 2	**Calculated Timeline Based on 2Peter 3:8 (years ago)**	**Biblical Account Reference Event**	**Scientific Determination of Time for Reference Event (plus citations)**
Beginning of Night One	5,110,000,000		
Beginning of Day One	4,745,000,000	'the earth'	The earth formed at about or slightly before 4.65 billion years ago.[4]
		'the deep' and 'the face of the waters'	Space ice supply predating the Sun[5]
		'the light'	The Sun forms at around 4.57 billion years ago.[6]
Beginning of Night Two	4,380,000,000		
Beginning of Day Two	4,015,000,000	'the sky'	Evidence exists for the atmosphere since the end of the Hadean Eon[7], which is about 4.0 billion years ago[8]. There is no evidence for an 'early' atmosphere.[9]
		'God…divided the waters which were	The hydrosphere existed by approximately 3.85

		under the expanse from the waters which were above the expanse'	billion years ago[10], and there is no direct evidence for a liquid water-based hydrosphere timing before 4.0 billion years ago.[11]
Beginning of Night Three	3,650,000,000		
Beginning of Day Three	3,285,000,000	'let the dry land be seen'	First landmass to appear above the waterline occurred at about 3.1 billion years ago[12]
		'green'	Divergence within land-colonizing bacteria between Cyanobacteria 'blue-green algae' and Chloroflexia estimated at 3.069 billion years ago (CI 2.922 to 3.184 billion years)[13]
		'herb sowing seed'	Estimated timing for divergence of endospore-forming bacteria phylum Firmicutes is about 3.054 billion years ago.[14]
Beginning of Night Four	2,920,000,000		
Beginning of Day Four	2,555,000,000	'for signs, and for seasons, and for days and years'	Sidereal Time-keeping possible since at least 2.45 billion years ago. Evidence of strong seasonality from 2.4 billion to 2.3 billion

			years ago during the Huronian Glaciation period[15]
Beginning of Night Five	2,190,000,000		
Beginning of Day Five	1,825,000,000		
Beginning of Night Six	1,460,000,000		

Chapter 22: DAY FIVE — The Waters Swarming with Swarms of Living Creatures as Sea Protozoa

I identify the 'the teeming living creature' in Genesis 1:20 as referring to the first sea protozoa. Protozoa are single-celled eukaryotic organisms. A eukaryote is defined as an organism consisting of a cell or cells in which the genetic material is DNA in the form of chromosomes contained within a distinct nucleus. Here is the relevant passage from Genesis Chapter One:

And God saith, `Let the waters teem with the teeming living creature, and fowl let fly on the earth on the face of the expanse of the heavens.' (Genesis 1:20, YLT)

In their article titled *Integrated genomic and fossil evidence illuminates life's early evolution and eukaryote origins* published in Nature Ecology & Evolution, a team of scientists from the University of Bristol, UK stated that "Modern eukaryotes emerged late in Earth history (<1.84 Ga), falsifying the hypothesis that the GOE facilitated their radiation."[16] GOE is an abbreviation for the 'Great Oxidation Event' that occurred approximately 2.5 to 2.3 billion years ago[97]. The team of scientists analyzed 29 protein-encoding genes from 102 species.

A Harvard professor and his colleagues implied that at about 1.8 billion years ago (1800 Myr) and later, well-preserved microfossil evidence was available to draw inferences from regarding eukaryotic organisms. Before I quote him, I want to provide some context. Myr means "million years." The term 'late Paleoproterozoic' probably means approximately 1.8 to 1.6 billion years ago, and the term 'early Mesoproterozoic' probably means approximately 1.6 billion years ago to 1.4 billion years ago. Dr. Andrew Knoll – a Harvard Professor Emeritus of Earth and Planetary Sciences, as well as a Harvard Professor of Biology – is an author of *Eukaryotic organisms in Proterozoic oceans* (also authored by E.J Javaux, D. Hewitt, and P. Cohen.) Professor Knoll and his colleagues wrote the following: "Steranes in 2780Myr shales from Western Australia have been interpreted as evidence that stem eukaryotes diverged early in Earth history (**Brocks *et al.* 1999**), but a general paucity of well-preserved microfossils precludes independent morphological inferences about eukaryotic biology prior to about 1800 Myr.

In summary, Late Palaeoproterozoic and Early Mesoproterozoic rocks preserve evidence for a moderate diversity of preservable eukaryotic organisms. This evidence includes cell walls without surface ornament (but with complex ultrastructure) and walls with regularly distributed surface ornamentation, with asymmetrically arranged processes that appear to reflect active cell growth, and with numerous symmetrically arranged processes."[17]

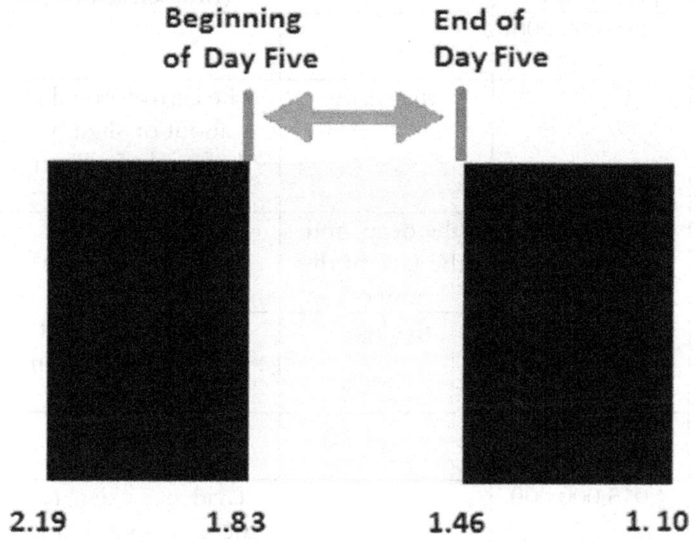

Times are given in Billions of Years and rounded to 3 Significant Figures

Day Five goes from 1,825,000,000 Julian years ago until 1,460,000,000 years ago.

Table 18: The Genesis Timeline, the Calculated Timeline Based on 2Peter 3:8, the Biblical Account's Reference Events, and the Scientific Determination of Time for Reference Events Thru the Waters Teeming with the Teeming Living Creatures in Chapter 22			
Timeline in Genesis 1 & 2	**Calculated Timeline Based on 2Peter 3:8 (years ago)**	**Biblical Account Reference Event**	**Scientific Determination of Time for Reference Event (plus citations)**
Beginning of Night One	5,110,000,000		
Beginning of Day One	4,745,000,000	'the earth'	The earth formed at about or slightly before 4.65 billion years ago.[4]
		'the deep' and 'the face of the waters'	Space ice supply predating the Sun[5]
		'the light'	The Sun forms at around 4.57 billion years ago.[6]
Beginning of Night Two	4,380,000,000		
Beginning of Day Two	4,015,000,000	'the sky'	Evidence exists for the atmosphere since the end of the Hadean Eon[7], which is about 4.0 billion years ago[8]. There is no evidence for an 'early' atmosphere.[9]
		'God...divided the waters which were under the expanse from the waters which were above the expanse'	The hydrosphere existed by approximately 3.85 billion years ago[10], and there is no direct evidence for a liquid water-based hydrosphere timing before 4.0 billion years ago.[11]

Beginning of Night Three	3,650,000,000		
Beginning of Day Three	3,285,000,000	'let the dry land be seen'	First landmass to appear above the waterline occurred at about 3.1 billion years ago[12]
		'green'	Divergence within land-colonizing bacteria between Cyanobacteria 'blue-green algae' and Chloroflexia estimated at 3.069 billion years ago (CI 2.922 to 3.184 billion years)[13]
		'herb sowing seed'	Estimated timing for divergence of endospore-forming bacteria phylum Firmicutes is about 3.054 billion years ago.[14]
Beginning of Night Four	2,920,000,000		
Beginning of Day Four	2,555,000,000	'and for seasons'	Evidence of strong seasonality from 2.4 billion to 2.3 billion years ago during the Huronian Glaciation period[15]
Beginning of Night Five	2,190,000,000		
Beginning of Day Five	1,825,000,000	'Let the waters teem with the teeming living creature'	Modern eukaryotes emerged late in Earth history (<1.84 Ga) [16], and late Palaeoproterozoic

			rocks preserve evidence for a moderate diversity of eukaryotic organisms.[17]
Beginning of Night Six	1,460,000,000		
Beginning of Day Six	1,095,000,000		
Beginning of Night Seven	730,000,000		

Chapter 23: DAY FIVE — The Living Creatures that Crawl in the Sea as Sea Sponges

In this chapter, I am identifying the 'living creature that crawls, with which the water swarms, according to their species' from Genesis 1:21 as a reference to sea sponges. Here is the relevant Day Five passage:

Then God said, "Let the waters swarm with swarms of living creatures! Let flying creatures fly above the land across the expanse of the sky." Then God created the large sea creatures and every living creature that crawls, with which the water swarms, according to their species, as well as every winged flying creature, according to their species. And God saw that it was good. Then God blessed them by saying, "Be fruitful and multiply and fill the water in the seas. Let the flying creatures multiply on the land." So there was evening and there was morning— a fifth day.

(Genesis 1:20-22, TLV) *(The Messianic Jewish Family Bible Society, 2015)*

It seems clear that to match the description in the passage, the creatures would be crawling, sea-dwelling multicellular animals.

A sea sponge is a type of sea-dwelling multicellular animal, which has varieties that can crawl up to 4 millimeters a day.[98] Regarding the timing, "Sponges diverged from other animals earlier in evolutionary history than any other known animal group, extant or extinct, with the first sponge-related record in earth history found in 1.8 billion year old sediments, based on a demosponge-specific chemofossil, 24-isopropylcholestane (McCaffrey *et al.*, 1994)."[18]

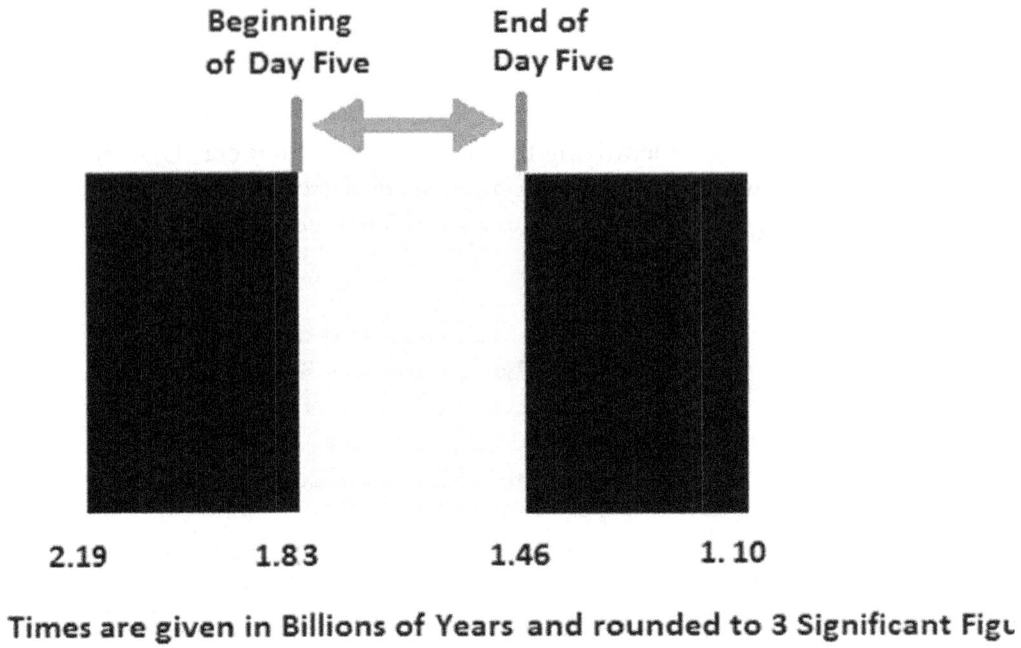

Day Five's time window spans from 1,825,000,000 years to 1,460,000,000 years ago. 1.8 billion years ago falls within the timespan for Day Five. The earliest reliable evidence for a sea sponge has a Day Five timing.

Table 19: The Genesis Timeline, the Calculated Timeline Based on 2Peter 3:8, the Biblical Account's Reference Events, and the Scientific Determination of Time for Reference Events Thru the Living Creatures that Crawl in the Sea in Chapter 23

Timeline in Genesis 1 & 2	Calculated Timeline Based on 2Peter 3:8 (years ago)	Biblical Account Reference Event	Scientific Determination of Time for Reference Event (plus citations)
Beginning of Night One	5,110,000,000		
Beginning of Day One	4,745,000,000	'the earth'	The earth formed at about or slightly before 4.65 billion years ago.[4]

		'the deep' and 'the face of the waters'	Space ice supply predating the Sun[5]
		'the light'	The Sun forms at around 4.57 billion years ago.[6]
Beginning of Night Two	4,380,000,000		
Beginning of Day Two	4,015,000,000	'the sky'	Evidence exists for the atmosphere since the end of the Hadean Eon[7], which is about 4.0 billion years ago[8]. There is no evidence for an 'early' atmosphere.[9]
		'God…divided the waters which were under the expanse from the waters which were above the expanse'	The hydrosphere existed by approximately 3.85 billion years ago[10], and there is no direct evidence for a liquid water-based hydrosphere timing before 4.0 billion years ago.[11]
Beginning of Night Three	3,650,000,000		
Beginning of Day Three	3,285,000,000	'let the dry land be seen'	First landmass to appear above the waterline occurred at about 3.1 billion years ago[12]
		'green'	Divergence within land-colonizing bacteria between Cyanobacteria 'blue-green algae' and Chloroflexia estimated at 3.069 billion years ago (CI 2.922 to 3.184 billion

			years)[13]
		'herb sowing seed'	Estimated timing for divergence of endospore-forming bacteria phylum Firmicutes is about 3.054 billion years ago.[14]
Beginning of Night Four	2,920,000,000		
Beginning of Day Four	2,555,000,000	'and for seasons'	Evidence of strong seasonality from 2.4 billion to 2.3 billion years ago during the Huronian Glaciation period[15]
Beginning of Night Five	2,190,000,000		
Beginning of Day Five	1,825,000,000	'Let the waters teem with the teeming living creature'	Modern eukaryotes emerged late in Earth history (<1.84 Ga) [16], and late Palaeoproterozoic rocks preserve evidence for a moderate diversity of eukaryotic organisms.[17]
		'living creature that crawls, with which the water swarms, according to their species'	Evidence of a chemical that is diagnostic for demosponges (a type of sea sponge) preserved in 1.8 billion year old sediments.[18]
Beginning of Night Six	1,460,000,000		

Beginning of Day Six	1,095,000,000		
Beginning of Night Seven	730,000,000		

Chapter 24: DAY FIVE — The Flying Creatures as Airborne Protozoa

The next event I will discuss is the appearance of airborne protozoa on Day Five. Below is the relevant passage.

Then God said, "Let the waters swarm with swarms of living creatures! Let flying creatures fly above the land across the expanse of the sky." Then God created the large sea creatures and every living creature that crawls, with which the water swarms, according to their species, as well as every winged flying creature, according to their species. And God saw that it was good. Then God blessed them by saying, "Be fruitful and multiply and fill the water in the seas. Let the flying creatures multiply on the land." So there was evening and there was morning—a fifth day.
(Genesis 1:20-22, TLV) [99]

Instead of 'birds' or 'fowl', the Tree of Life Version (TLV) Bible translation uses the more general 'flying creatures' with the word 'winged'. It is noteworthy that the symbol translated as 'winged' is kānāp̄, which corresponds to Strong's symbol H3671 and can be defined as meaning 'sort' and 'flying' which are also listed under Strong's Definitions for the symbol on the Blue Letter Bible's website[100]. Supposing that flying or 'airborne' is equivalent (…every airborne flying creature, according to their species) or that 'sort' is intended (…every sort [of] flying creature, according to their species…)[101], this would open up another possibility that would have likely been overlooked in the past: the first airborne protozoa.

A reader might express concern if the symbol within the Biblical narrative could only be interpreted as 'birds' as that would imply that birds would be appearing out of sequence in the Biblical narrative – birds should appear after the first land animals. If airborne protozoa did not exist or if their estimated divergence timing was not a Day Five timing, a misunderstanding here would have implied contradictions between science and the Biblical narrative. However, airborne protozoa do exist, and the estimated divergence timings for some of their species is during Day Five.

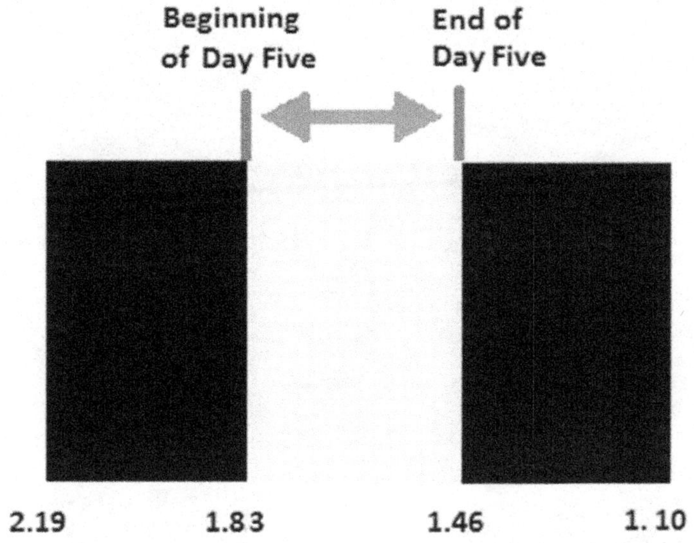

Times are given in Billions of Years and rounded to 3 Significant Figures

Day Five spans from about 1,825,000,000 Julian years ago up to about 1,460,000,000 years ago.

Since most readers may be unfamiliar with airborne protozoa, providing some background information seems appropriate. The term protozoa means 'first animals' and for these small creatures, "all life functions are carried out within the confines of a single cell."[102] Harold E. Schlichting Jr., a botany professor from North Carolina State University, compiled a summation of viable airborne algae and protozoa in his 1969 The Importance of Airborne Algae and Protozoa.[103]

In his 1973 work The Effects of Some Air Pollutants and Meteorological Conditions on Airborne Algae and Protozoa, Paul E. Smith from the Westinghouse Electric Corporation's Environmental Systems Center stated that "19 species of protozoa were collected from the atmosphere…" during his experiment.[104]

Below is a list of protozoan-species data from Smith's 1973 journal article for protozoa cultured from the atmosphere.

Protozoa listed in Smith's Table II.[105]
Anisonema acinus
Bodo edax
Chryptochrysis commutata
Chrysapsis sagene
Coccomonas sp.
Colpoda cucullus
Entosiphon sp.
Euplotes patella
Kahlia sp.
Monas elongata
Monas guttula
Myxochrysis sp.
Paramecium trichium
Peranema sp.
Petalomonas sp.
Pseudoprorodon sp.
Tetrahymena sp.
Trichospira sp.
Unidentified zooflagellate

Britannica, famous for its encyclopedia, includes in its Protozoan article under the subheading, 'Adaptations', that "Many protozoans respond to adverse environmental conditions by encysting: they secrete a thick, tough wall around themselves and effectively enter a quiescent state comparable to hibernation. The ability to form a resistant cyst is widespread among diverse protistan groups and probably developed early in their evolutionary history. Resting cysts also are easily carried by the wind and form an important means of dispersal for species that live in the soil or are common in ephemeral ponds and pools."[106]

I looked at other scientists who also studied protozoans in the air. Scientists Savvas Genitsaris and Maria Moustaka-Gouni from the Aristotle University of Thessaloniki's School of Biology, and Konstantinos A. Kormas from the University of Thessaly's School of Agricultural Sciences studied airborne microeukaryotes in experimental water containers. They listed

Amoeba cf. proteus as having a moderate frequency of occurrence as an airborne microeukaryote found in the water containers, and Amoeba spp. having a high frequency of occurrence[107]. The abbreviation 'spp.' is a Latin abbreviation for multiple species – thus they found multiple Amoeba species in the water containers. Amoebozoa is noted on Wikipedia as a phylum containing amoeboid protists, the amoebas.

Dr. Joseph F. Saunders, the program scientist for biosatellite flights and chief of environmental biology at the office of space science application and manned space flight at NASA[108], edited a work by NASA that noted the rates of growth of Amoeba from the Biosatellite II flight experiment as compared to Earth controls[109]. It was noted for the 'giant multinucleate amoeba *Pelomyxa carolinensis*' that "…this organism appears to be independent of gravity on Earth…"[110] In the Discussion section with the subheading 'Survival and Growth of Amoebae in Space', it was written that, "Amoebae flown in space not only survived the total flight profile but fed, assimilated food, and grew in the weightless environment."[111]

In 2011, Scientists Laura Wegener Parfrey, Daniel J.G. Lahr, Andrew H. Knoll, and Laura A. Katz published in the Proceedings of the National Academy of Sciences of the United States of America their work *Estimating the timing of early eukaryotic diversification with multigene molecular clocks*. Timetree.org cited Parfrey et al as having estimated that Amoebozoa, the phylum containing amoeboid protists, the amoebas, came into existence 1.5383 billion years ago. According to timetree.org, Parfrey et al listed a grouping spanning from Acandthamoeba castellanii to Dictyostelium discoideum in the published tree's data with the time equal to 1.538 billion years ago.[112] Parfrey et al estimate of the root of Amoebozoa having a 95% highest-probability density, spanning a significant time range of time, most of which (but not all) falling on Day Five.

Timetree.org estimates Amoebozoa's divergence at 1.522 billion years ago.[113]

Table 20: The Genesis Timeline, the Calculated Timeline Based on 2Peter 3:8, the Biblical Account's Reference Events, and the Scientific Determination of Time for Reference Events Thru the Flying Creatures in Chapter 24			
Timeline in Genesis 1 & 2	**Calculated Timeline Based on 2Peter 3:8 (years ago)**	**Biblical Account Reference Event**	**Scientific Determination of Time for Reference Event (plus citations)**
Beginning of Night One	5,110,000,000		
Beginning of Day One	4,745,000,000	'the earth'	The earth formed at about or slightly before 4.65 billion years ago.[4]
		'the deep' and 'the face of the waters'	Space ice supply predating the Sun[5]
		'the light'	The Sun forms at around 4.57 billion years ago.[6]
Beginning of Night Two	4,380,000,000		
Beginning of Day Two	4,015,000,000	'the sky'	Evidence exists for the atmosphere since the end of the Hadean Eon[7], which is about 4.0 billion years ago[8]. There is no evidence for an 'early' atmosphere.[9]
		'God…divided the waters which were under the expanse from the waters which were above the expanse'	The hydrosphere existed by approximately 3.85 billion years ago[10], and there is no direct evidence for a liquid water-based hydrosphere timing before 4.0 billion years ago.[11]

Beginning of Night Three	3,650,000,000		
Beginning of Day Three	3,285,000,000	'let the dry land be seen'	First landmass to appear above the waterline occurred at about 3.1 billion years ago[12]
		'green'	Divergence within land-colonizing bacteria between Cyanobacteria 'blue-green algae' and Chloroflexia estimated at 3.069 billion years ago (CI 2.922 to 3.184 billion years)[13]
		'herb sowing seed'	Estimated timing for divergence of endospore-forming bacteria phylum Firmicutes is about 3.054 billion years ago.[14]
Beginning of Night Four	2,920,000,000		
Beginning of Day Four	2,555,000,000	'and for seasons'	Evidence of strong seasonality from 2.4 billion to 2.3 billion years ago during the Huronian Glaciation period[15]
Beginning of Night Five	2,190,000,000		
Beginning of Day Five	1,825,000,000	'Let the waters teem with the teeming living creature'	Modern eukaryotes emerged late in Earth history (<1.84 Ga)[16], and late Palaeoproterozoic

			rocks preserve evidence for a moderate diversity of eukaryotic organisms.[17]
		'living creature that crawls, with which the water swarms, according to their species'	Evidence of chemical that is diagnostic for demosponges (a type of sea sponge) preserved in 1.8 billion year old sediments.[18]
		'every winged flying creature, according to their species'	With multiple amoebas noted as being present in atmospheric air[114], the grouping that contains amoebas, Amoebozoa, is estimated to have come into existence is 1.538 billion years ago.[115]
Beginning of Night Six	1,460,000,000		
Beginning of Day Six	1,095,000,000		
Beginning of Night Seven	730,000,000		

Chapter 25: DAY SIX — Animal of the earth as the Earthworm

The next event I will discuss is the presence of earthworms on Day Six as the 'beast of the earth' i.e. the soil. Below is the Biblical passage.

And God saith, `Let the earth bring forth the living creature after its kind, cattle and creeping thing, and beast of the earth after its kind:' and it is so. 25 And God maketh the beast of the earth after its kind, and the cattle after their kind, and every creeping thing of the ground after its kind, and God seeth that it is good. (Genesis 1: 24-25, YLT)

There is evidence of worm burrows in sandstone in India that are from more than 1.0 billion years old up to as early as 1.1 billion years ago.[21] The corresponding article titled *Triploblastic Animals More Than 1 Billion Years Ago: Trace Fossil Evidence from India* from the scientists Adolf Seilacher, Pradip K. Bose, and Friedrich Pflüger about these animals was published in *Science* magazine in 1998.[116]

Day Six begins at about 1.095 billion years ago and goes to 730 million years ago.

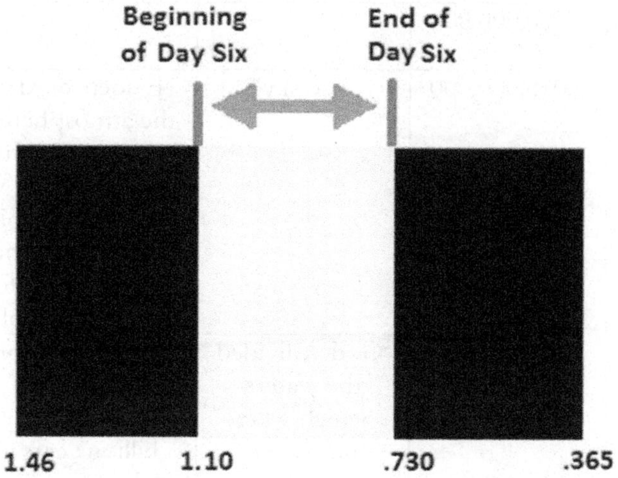

Times are given in Billions of Years and rounded to 3 Significant Figures

The worm burrows' timing is consistent with a Day Six timing.

Table 21: The Genesis Timeline, the Calculated Timeline Based on 2Peter 3:8, the Biblical Account's Reference Events, and the Scientific Determination of Time for Reference Events Thru the beast of the Earth in Chapter 25			
Timeline in Genesis 1 & 2	**Calculated Timeline Based on 2Peter 3:8 (years ago)**	**Biblical Account Reference Event**	**Scientific Determination of Time for Reference Event (plus citations)**
Beginning of Night One	5,110,000,000		
Beginning of Day One	4,745,000,000	'the earth'	The earth formed at about or slightly before 4.65 billion years ago.[4]
		'the deep' and 'the face of the waters'	Space ice supply predating the Sun[5]
		'the light'	The Sun forms at around 4.57 billion years ago.[6]

Beginning of Night Two	4,380,000,000		
Beginning of Day Two	4,015,000,000	'the sky'	Evidence exists for the atmosphere since the end of the Hadean Eon[7], which is about 4.0 billion years ago[8]. There is no evidence for an 'early' atmosphere.[9]
		'God…divided the waters which were under the expanse from the waters which were above the expanse'	The hydrosphere existed by approximately 3.85 billion years ago[10], and there is no direct evidence for a liquid water-based hydrosphere timing before 4.0 billion years ago.[11]
Beginning of Night Three	3,650,000,000		
Beginning of Day Three	3,285,000,000	'let the dry land be seen'	First landmass to appear above the waterline occurred at about 3.1 billion years ago[12]
		'green'	Divergence within land-colonizing bacteria between Cyanobacteria 'blue-green algae' and Chloroflexia estimated at 3.069 billion years ago (CI 2.922 to 3.184 billion years)[13]

		'herb sowing seed'	Estimated timing for divergence of endospore-forming bacteria phylum Firmicutes is about 3.054 billion years ago.[14]
Beginning of Night Four	2,920,000,000		
Beginning of Day Four	2,555,000,000	'and for seasons'	Evidence of strong seasonality from 2.4 billion to 2.3 billion years ago during the Huronian Glaciation period[15]
Beginning of Night Five	2,190,000,000		
Beginning of Day Five	1,825,000,000	'Let the waters teem with the teeming living creature'	Modern eukaryotes emerged late in Earth history (<1.84 Ga)[16], and late Palaeoproterozoic rocks preserve evidence for a moderate diversity of eukaryotic organisms.[17]
		'living creature that crawls, with which the water swarms, according to their species'	Evidence of chemical that is diagnostic for demosponges (a type of sea sponge) preserved in 1.8 billion year old sediments.[18]
		'every winged flying creature, according to their species'	With multiple amoebas noted as being present in atmospheric air[117], the grouping that

			contains amoebas, Amoebozoa, is estimated to have come into existence is 1.538 billion years ago.[118]
Beginning of Night Six	1,460,000,000		
Beginning of Day Six	1,095,000,000	'beast of the earth'	Worm burrows in sandstone in India that are from more than 1.0 billion years old up to as early as 1.1 billion years ago.[21]
Beginning of Night Seven	730,000,000		
Beginning of Day Seven	365,000,000		
End of Day Seven	0		

Chapter 26: DAY SIX — Addressing the Perceived Contradiction of the Livestock

I will discuss the livestock, which is referenced in the Day Six passage. Below is the Biblical passage that describes the creation of the livestock.

God said, "Let the earth bring forth living creatures after their kind, livestock, creeping things, and animals of the earth after their kind;" and it was so. God made the animals of the earth after their kind, and the livestock after their kind, and everything that creeps on the ground after its kind. God saw that it was good. (Genesis 1:24-25, New Heart English Bible)

Some biologists have considered jawed land animals with a spinal cord to have descended from a jawed fish.[119] If true, this narrative would imply that the jawed fish is the ancestor of all terrestrial jawed chordates and would require that mammals be no older than the time when the jawed fish first came into existence. However, this is and has been for decades (if not longer) the source of considerable controversy among biologists. Biologist Kevin J Peterson puts it this way: "The origin of the Craniata (hagfish + Vertebrata [Vertebrata = lamprey + Gnathostoma]—Janvier, 1981), one of the three subphyla of the Phylum Chordata, has generated more controversy in terms of numbers of likely ancestors than the origin of any other metazoan group."[120]

Without resolution and consensus on the origins of the first vertebrates with absolute orderings, I can't claim that modern Science has rendered absolute and reliable times for the orderings for land animals – such as the livestock.

Day Six spans from 1,095,000,000 years ago until about 730,000,000 years ago.

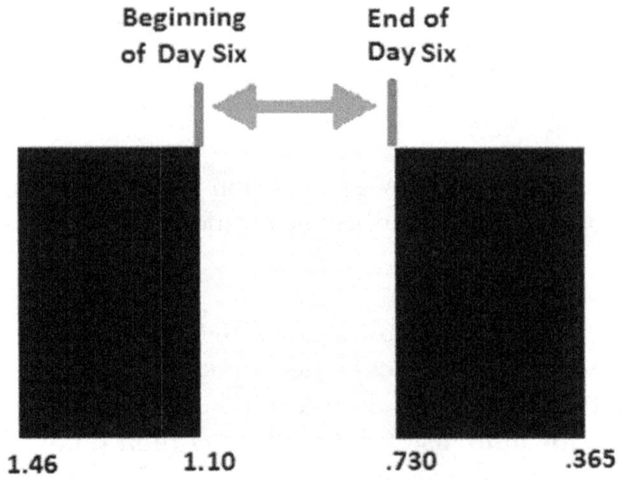

Times are given in Billions of Years and rounded to 3 Significant Figures

Chapter 27: DAY SIX — Man and Woman

Below is the Genesis passage that places the creation of both a man and a woman on Day Six.

God said, "Let us make man in our image, after our likeness: and let them have dominion over the fish of the sea, and over the birds of the sky, and over the livestock, and over all the earth, and over every creeping thing that creeps on the earth." God created man in his own image. In God's image he created him; male and female he created them. God blessed them. God said to them, "Be fruitful, multiply, fill the earth, and subdue it. Have dominion over the fish of the sea, over the birds of the sky, and over every living thing that moves on the earth." God said, "Behold, I have given you every herb yielding seed, which is on the surface of all the earth, and every tree, which bears fruit yielding seed. It will be your food. To every animal of the earth, and to every bird of the sky, and to everything that creeps on the earth, in which there is life, I have given every green herb for food;" and it was so.

God saw everything that he had made, and, behold, it was very good. There was evening and there was morning, a sixth day. (Genesis 1: 26-31, NHEB, footnotes removed)

It is beyond my ability to demonstrate the creation of the first Man and first Woman as detailed in the Scriptures on purely secular scientific grounds. Indeed, such a demonstration – if one is able to demonstrate it on the secular grounds of Science -- would be more than worthy of a book of its own.

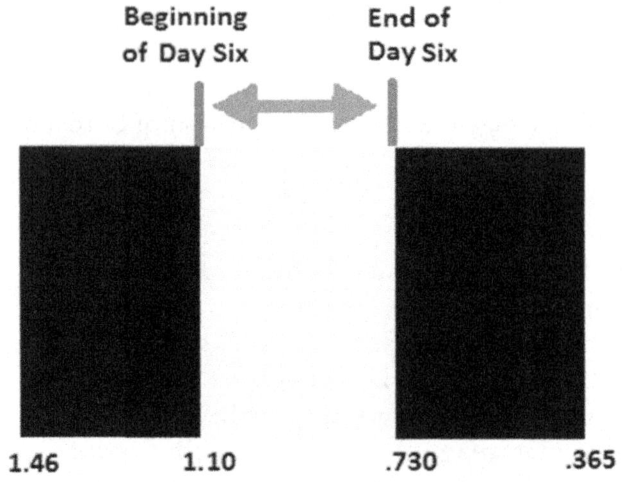

Times are given in Billions of Years and rounded to 3 Significant Figures

Chapter 28: DAY SIX — Land Plants

The Bible requires land plants to fall on Day Six according to the text in the first two chapters of Genesis. Below are the relevant quotes:

And God saith, 'Let Us make man in Our image, according to Our likeness, and let them rule over fish of the sea, and over fowl of the heavens, and over cattle, and over all the earth, and over every creeping thing that is creeping on the earth.' ***And God prepareth the man in His image; in the image of God He prepared him, a male and a female He prepared them. And God blesseth them, and God saith to them, 'Be fruitful, and multiply, and fill the earth, and subdue it, and rule over fish of the sea, and over fowl of the heavens, and over every living thing that is creeping upon the earth.' And God saith, 'Lo, I have given to you every herb sowing seed, which*** **is** ***upon the face of all the earth, and every tree in which*** **is** ***the fruit of a tree sowing seed, to you it is for food; and to every beast of the earth, and to every fowl of the heavens, and to every creeping thing on the earth, in which*** **is** ***breath of life, every green herb*** **is** ***for food:' and it is so. And God seeth all that He hath done, and lo, very good; and there is an evening, and there is a morning -- day the sixth. (Genesis 1:26-31, YLT)***

And from the second chapter of Genesis:

And no plant of the field was yet in the earth, and no

herb of the field had yet sprung up; for Jehovah God had not caused it to rain upon the earth: and there was not a man to till the ground; but there went up a mist from the earth, and watered the whole face of the ground. And Jehovah God formed man of the dust of the ground, and breathed into his nostrils the breath of life; and man became a living soul. And Jehovah God planted a garden eastward, in Eden; and there he put the man whom he had formed. And out of the ground made Jehovah God to grow every tree that is pleasant to the sight, and good for food; the tree of life also in the midst of the garden, and the tree of the knowledge of good and evil. (Genesis 2:5-9, ASV)

Later in Genesis Chapter Two, the first woman, Eve is created.

...and the rib, which Jehovah God had taken from the man, made he a woman, and brought her unto the man. (Genesis 2:22, ASV)

The order is important. Since Adam and Eve were both created in Genesis Chapter One during Day Six, and according to the text in Genesis chapter Two, the order goes Man, then land plants, then Woman, the text maintains that land plants is also a Day Six event. A rigorous reading of the text forces land plants to be required to fall on Day Six according to the Bible because land plants are between two other Day Six events: the creation of Man and the creation of Woman.

Below is a reminder of the window of time that Day Six encompasses. Day Six goes from 1.095 billion years ago to 730 million years ago.

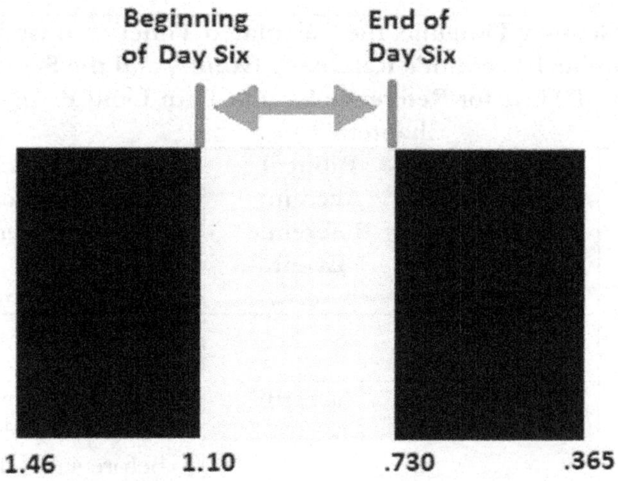

Times are given in Billions of Years and rounded to 3 Significant Figures

The scientific name for land plants is 'Embryophyta' or the embryophytes, and according to a team of scientists having representation drawn from Nanjing Normal University, from South China Agricultural University in Guangzhou, and from Temple University in Philadelphia: "Taken together, our analyses support that crown embryophytes originated in the Neoproterozoic (fig. 4, Strategy 2: 980–682 Ma.)"[22] Of the 298 million year range that the scientists provided, about 250 million years occur during the Day Six time window, and a span of about 48 million years would fall on Night Seven.

The majority of the time range falls within Day Six, which supports a Day Six timing for this event.

Table 22: The Genesis Timeline, the Calculated Timeline Based on 2Peter 3:8, the Biblical Account's Reference Events, and the Scientific Determination of Time for Reference Events Thru Land Plants in Chapter 28			
Timeline in Genesis 1 & 2	**Calculated Timeline Based on 2Peter 3:8 (years ago)**	**Biblical Account Reference Event**	**Scientific Determination of Time for Reference Event (plus citations)**
Beginning of Night One	5,110,000,000		
Beginning of Day One	4,745,000,000	'the earth'	The earth formed at about or slightly before 4.65 billion years ago.[4]
		'the deep' and 'the face of the waters'	Space ice supply predating the Sun[5]
		'the light'	The Sun forms at around 4.57 billion years ago.[6]
Beginning of Night Two	4,380,000,000		
Beginning of Day Two	4,015,000,000	'the sky'	Evidence exists for the atmosphere since the end of the Hadean Eon[7], which is about 4.0 billion years ago[8]. There is no evidence for an 'early' atmosphere.[9]
		'God…divided the waters which were under the expanse from the waters which were above the expanse'	The hydrosphere existed by approximately 3.85 billion years ago[10], and there is no direct evidence for a liquid water-based hydrosphere timing before 4.0 billion years ago.[11]

Beginning of Night Three	3,650,000,000		
Beginning of Day Three	3,285,000,000	'let the dry land be seen'	First landmass to appear above the waterline occurred at about 3.1 billion years ago[12]
		'green'	Divergence within land-colonizing bacteria between Cyanobacteria 'blue-green algae' and Chloroflexia estimated at 3.069 billion years ago (CI 2.922 to 3.184 billion years)[13]
		'herb sowing seed'	Estimated timing for divergence of endospore-forming bacteria phylum Firmicutes is about 3.054 billion years ago.[14]
Beginning of Night Four	2,920,000,000		
Beginning of Day Four	2,555,000,000	'and for seasons'	Evidence of strong seasonality from 2.4 billion to 2.3 billion years ago during the Huronian Glaciation period[15]
Beginning of Night Five	2,190,000,000		
Beginning of Day Five	1,825,000,000	'Let the waters teem with the teeming living creature'	Modern eukaryotes emerged late in Earth history (<1.84 Ga) [16], and late Palaeoproterozoic

			rocks preserve evidence for a moderate diversity of eukaryotic organisms.[17]
		'living creature that crawls, with which the water swarms, according to their species'	Evidence of chemical that is diagnostic for demosponges (a type of sea sponge) preserved in 1.8 billion year old sediments.[18]
		'every winged flying creature, according to their species'	With multiple amoebas noted as being present in atmospheric air[121], the grouping that contains amoebas, Amoebozoa, is estimated to have come into existence is 1.538 billion years ago.[122]
Beginning of Night Six	1,460,000,000		
Beginning of Day Six	1,095,000,000	'beast of the earth'	Worm burrows in sandstone in India that are from more than 1.0 billion years old up to as early as 1.1 billion years ago.[21]
		'every tree in which is the fruit of a tree sowing seed'	Scientific analysis supported that crown embryophytes originated sometime between 980 and 682 million years ago.[22]
Beginning of Night Seven	730,000,000		
Beginning of Day Seven	365,000,000		
End of Day Seven	0		

Chapter 29: DAY SEVEN — The Heavens Completed as No New Galaxies Formed During the Day of Rest

By the beginning of the seventh day in chapter two in Genesis, God had ceased from his work according to the text.

And the heavens and the earth are completed, and all their host;

and God completeth by the seventh day His work which He hath made, and ceaseth by the seventh day from all His work which He hath made.

And God blesseth the seventh day, and sanctifieth it, for in it He hath ceased from all His work which God had prepared for making. (Genesis 2: 1-3, YLT)

This gentle sounding passage contains an astonishing test and result: "…the heavens and the earth are completed, and all their host;" Let us suppose that, "…the heavens…are completed;" is a roughly translated reference to the end of the formation of galaxies in the observable universe, and 'the earth' is a reference to our planet. A galaxy is "a large system of stars held together by mutual gravitation and isolation from similar systems by vast regions of space." If this statement is about the end of the formation of galaxies, it is truly an astonishing test by science of the Bible's accuracy. David J. Eicher, editor of Astronomy, wrote that "By using the Hubble Space Telescope, astronomers have estimated that some 100 billion galaxies must exist in the cosmos. And the number may be much greater than that. Probably some 2 trillion galaxies existed in the early universe, but it seems clear that galaxies near each other are drawn together by gravity and combine over cosmic time."[123] Day Seven began about 365 million years ago and under my model continued to the end of the model (year zero). Did any galaxy form during Day Seven?

The answer is 'No.' This answer is in accord with scholarly sources. A. I. Terlevich and Duncan A. Forbes made a catalogue of galaxy ages and analyzed local galaxy ages and metallicities in their work titled *A catalogue and analysis of local galaxy ages and metallicities*, which was published in the Monthly Notices of the Royal Astronomical Society in 2002. In their Table 3, they "…listed age estimates for young or proto-ellipticals that we are aware of in the literature. These estimates come from stellar dynamics, proto-globular clusters, extended structures etc."[124] The young galaxies matching their desired criteria included a dozen galaxies: NGC 1052, NGC 1700, NGC 2865, NGC 3156, NGC 3610, NGC 3656, NGC 3921, NGC 3921, NGC 5322, NGC 6776, NGC 7252, NGC 7585, and ESO341-IG04. The range of ages of the twelve galaxies has the youngest galaxy's specified age as being 0.7 billion years. The age of galaxy NGC 2865 was given by the authors as less than 1.5 billion years, but prior work by Forbes[125] in the same table labels this galaxy as being 1.2 billion years old. Galaxy NGC 2865 was assigned an age of 1.8 ± 0.5 billion years old by Rampazzo et al in their work *The Galaxy Evolution Explorer UV emission in shell galaxies: tracing galaxy 'rejuvenation' episodes*, which was published in Monthly Notices of the Royal Astronomical Society.[126] For the galaxy NGC 7252, which had the youngest specified age listed in Table 3, Terlevich and Forbes cited Forbes's earlier work as having assigned it an age of 0.7 billion years.

I decided to examine the galaxy with the youngest specified age: galaxy NGC 7252. Francois Schweitzer and Patrick Seitzer's *Ages and Metallicities of Young Globular Clusters in the Merger Remnant NGC 7252* in The Astronomical Journal states that the galaxy "NGC 7252 itself shows the characteristics of a 0.5--1 Gyr old protoelliptical galaxy…"[127] and "Both the observations and the model strongly suggest that NGC 7252 is a 0.5--1 Gyr old protoelliptical."[128] This lower estimate of age is in accord with educational materials produced by NASA's website for children also state that "The newest galaxy we know of formed only about 500 million years ago."[129] The educational materials I am referring to were lasted updated in 2019.[130]

The galaxy that was thought to be the youngest galaxy known by astronomers was the galaxy I Zwicky 18.[131] I Zwicky 18 was estimated to have an age of about 500 million years. However, 1 Zwicky 18 was eventually found not to be as young as the astronomers had estimated. Despite the estimated presence of more than 100 billion galaxies in the observable universe -- as far as I know from going through NASA's website and other materials – none of them were created within the last 365 million years. No new galaxies on Day Seven.

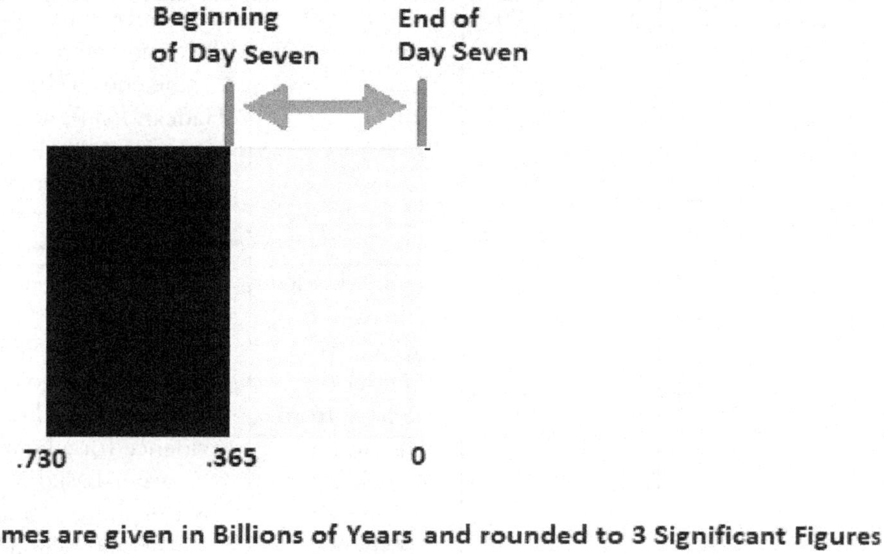

Table 23: The Genesis Timeline, the Calculated Timeline Based on 2Peter 3:8, the Biblical Account's Reference Events, and the Scientific Determination of Time for Reference Events Thru the Heavens and Their Host being Completed and God Resting on Day Seven implying No New Galaxies during Day Seven in Chapter 29			
Timeline in Genesis 1 & 2	**Calculated Timeline Based on 2Peter 3:8 (years ago)**	**Biblical Account Reference Event**	**Scientific Determination of Time for Reference Event (plus citations)**
Beginning of Night One	5,110,000,000		
Beginning of Day One	4,745,000,000	'the earth'	The earth formed at about or slightly before 4.65 billion years ago.[4]
		'the deep' and 'the face of the waters'	Space ice supply predating the Sun[5]
		'the light'	The Sun forms at around 4.57 billion

			years ago.[6]
Beginning of Night Two	4,380,000,000		
Beginning of Day Two	4,015,000,000	'the sky'	Evidence exists for the atmosphere since the end of the Hadean Eon[7], which is about 4.0 billion years ago[8]. There is no evidence for an 'early' atmosphere.[9]
		'God…divided the waters which were under the expanse from the waters which were above the expanse'	The hydrosphere existed by approximately 3.85 billion years ago[10], and there is no direct evidence for a liquid water-based hydrosphere timing before 4.0 billion years ago.[11]
Beginning of Night Three	3,650,000,000		
Beginning of Day Three	3,285,000,000	'let the dry land be seen'	First landmass to appear above the waterline occurred at about 3.1 billion years ago[12]
		'green'	Divergence within land-colonizing bacteria between Cyanobacteria 'blue-green algae' and Chloroflexia estimated at 3.069 billion years ago (CI 2.922 to 3.184 billion years) [13]

		'herb sowing seed'	Estimated timing for divergence of endospore-forming bacteria phylum Firmicutes is about 3.054 billion years ago.[14]
Beginning of Night Four	2,920,000,000		
Beginning of Day Four	2,555,000,000	'and for seasons'	Evidence of strong seasonality from 2.4 billion to 2.3 billion years ago during the Huronian Glaciation period[15]
Beginning of Night Five	2,190,000,000		
Beginning of Day Five	1,825,000,000	'Let the waters teem with the teeming living creature'	Modern eukaryotes emerged late in Earth history (<1.84 Ga) [16], and late Palaeoproterozoic rocks preserve evidence for a moderate diversity of eukaryotic organisms.[17]
		'living creature that crawls, with which the water swarms, according to their species'	Evidence of chemical that is diagnostic for demosponges (a type of sea sponge) preserved in 1.8 billion year old sediments.[18]
		'every winged flying creature, according to their species'	With multiple amoebas noted as being present in atmospheric air[132], the grouping that

			contains amoebas, Amoebozoa, is estimated to have come into existence is 1.538 billion years ago.[133]
Beginning of Night Six	1,460,000,000		
Beginning of Day Six	1,095,000,000	'beast of the earth'	Worm burrows in sandstone in India that are from more than 1.0 billion years old up to as early as 1.1 billion years ago.[21]
		'every tree in which is the fruit of a tree sowing seed'	Scientific analysis supported that crown embryophytes originated sometime between 980 and 682 million years ago.[22]
Beginning of Night Seven	730,000,000		The newest known galaxy according to NASA[23], which formed about 500 million years ago.
Beginning of Day Seven	365,000,000	'the heavens…completed, …and God…ceaseth by the seventh day from all His work which He hath made'	No known galaxy has formed within the last 365 million years.
End of Day Seven	0		

Chapter 30: Chance that the Genesis Writer would Provide the Correct Order for Each Event for Day One thru Day Six

With the correct ordering and the correct times given for all events of known timings, the question of 'How did the Genesis Writer and Saint Peter know?" is fair. The Genesis Writer gave the ordering of the events while Saint Peter's statement effectively provided times for each night and day in the Genesis Creation Account. Both feats were done over 1800 years ago at a time devoid of modern science. How did they know?

If by divine knowledge, then there's a God. In case you are wondering if they just 'got lucky', I want to go over the chance of this happening, which can be measured. I will supply the reasoning for calculating this probability, some of its premises, and include thirteen events in the calculation.

Below is a recap of the events of known timing that I covered earlier.

1. The Earth and its Age.

2. The Sun and its Age.

3. The Atmosphere and its Age

4. Oceans and their Age

5. Dry Land and its Age

6. Green and its Age

7. Endospores and their Age

8. Seasons and their Age

9. Swarms of Swarming Living Creatures (Marine Protozoa) and their Age

10. Crawling Ocean Creatures (Sponges) and their Age.

11. Airborne Microscopic Creatures (Airborne Protozoa) and their Age.

12. Earthworms and their Age.

13. Land Plants and their Age.

While I covered some other evidence that was supportive of the Biblical narrative, here are 13 events that are external to the Bible, yet their ordering and timing is consistent with the Biblical account. I will cover the probability of a new galaxy not forming on Day Seven separately from these 13 events.

Suppose that we are not in a world with the Divine, what is the chance that the Genesis Writer took a guess and lined the 13 events up correctly by chance?

In this question, there is only 1 correct arrangement of events, and the probability relies on placing the number of correct arrangements over the number of possible arrangements, that is the total number of arrangements. The total number of arrangements is 6,227,020,800. The total number of arrangements is equal to 13 x 12 x 11 x 10 x 9 x 8 x 7 x 6 x 5 x 4 x 3 x 2 x 1, which is equal to 6,227,020,800. This is because for placing 13 events, one can view it as 13 slots to put events in as all events have to be placed. So for the 1st slot to be filled, there are 13 events to choose from. After placing the first event, it is checked off the list and not placed again. For the next slot, there are 12 events to choose from, exactly 1 of which will be placed there. After placing the 2nd event, it is checked off the list as well (11 events remain to be placed). For the next slot, there are 11 events to choose from to put in that particular slot, 1 of which can be placed there. After placing the 3rd event, it is checked off the list as well and then only 10 events remain. One continues in this manner, until the last event is to be placed, and there is only 1 slot remaining, hence the 'x 1' at the end. This is a 1 in 6,227,020,800 shot. For those who like math, this is 1 over 13 factorial.

So there's a 1 in 6,227,020,800 shot that the Genesis Writer would correctly arrange the <u>relative order</u> of the 13 events if he just made guesses.

Chapter 31: Chance that the Genesis Writer and Saint Peter would Provide the Correct Time for Each Event in Day One thru Day Six

If Saint Peter was divinely inspired and spoke about it, the chance of him giving the correct answer is 100%. If we suppose that Saint Peter was not divinely inspired, then what is the chance of him giving the correct times for the 13 events? I will assume that any particular year was as likely as any other year as far as guesses go. That is, I will use a uniform distribution for Saint Peter's supposedly dishonest guessing – one year is as likely as another.

It's important to realize that Saint Peter's statement provided both times for the individual events as well as a 'start time' for the narrative (start time = 14 x 1 day of the Lord's Time = 5.11 billion years). Supposing that the start time is trivial for him to know (it is not), the chance of placing any unknown event on the correct slot with 14 slots to choose from is 1 in 14. 14 slots to choose from -- all other things being equal – gives a 1 in 14 chance per event. So with 13 events, this is equal to the product of 1/14 multiplied out with 13 entries. This is equal to 1 over 14 x 14 x 14 x 14 x 14 x 14 x 14 x 14 x 14 x 14 x 14 x 14 x 14 = 1 over 793,714,773,254,144. St. Peter has a little worse than a 1 in a 793 trillion shot to guess the correct timing for 13 events.

Chapter 32: Chance that the Genesis Writer and Saint Peter would both be correct without prior Knowledge (without Divine Knowledge)

Many people – non-scholarly and scholarly alike, coming from a wide variety of theological and philosophical positions – have insisted that the scientific evidence must be weighed on this question, "Is there a God?"

Under very conservative assumptions, the overall probability of St. Peter guessing along with a Genesis writer guessing and getting the time and order as determined by modern science correct by chance would be:

Probability of Guessing Correctly = (Pass Relative Order Test) x (Pass Correct Time Window Test)

= (1 / number of events factorial) x (chance that event lands on the correct day)$^{\text{number of events}}$

= (1 / (13 x 12 x 11 x 10 x 9 x 8 x 7 x 6 x 5 x 4 x 3 x 2 x 1)) x (1 / 14)^13

= (1 / 6,227,020,800) x (1 / 793,714,773,254,144)

= 1 / 4,942,478,402,320,838,374,195,200

The probability of both the Genesis Writer and Saint Peter guessing correctly is 1 in 4,942,478,402,320,838,374,195,200. The chance of them correctly guessing the scientific placement with the correct relative order and the correct timing for the aforementioned 13 events is a less than 1 in a 4 septillion-shot.

Chapter 33: Examination of the Chance that the Genesis Writer and Saint Peter would Pass the No-New-Galaxies-During-Day-Seven Test

As recorded in Chapter 28, according to scientists there are at least 100 billion galaxies in the observable universe. NASA official Dr. Edward J. Wollack wrote on a NASA web page with regards to the Wilkinson Microwave Anisotrophy Probe satellite, which is called WMAP, that, "As of 2013, WMAP determined these parameters with an accuracy of better than than 1.5%. In turn, knowing the composition with this precision, we can estimate the age of the universe to about 0.4%: 13.77 ± 0.059 billion years!"[134] The universe is about 13.8 billion years old, and its age is known with precision.

Day Seven is 365 million years long. The observable universe's age is estimated to be about 13.8 billion years old. 13.8 billion years is equal to 13800 million years. 365 million years divided by 13800 million years is equal to approximately .0264497, or about 2.6 % of the total time of the universe's existence. Calculating the improbability of no new galaxies occurring on Day Seven depends on knowing the age distribution of the 100 billion galaxies that NASA has had the opportunity to examine. NASA's teaching materials for children mention that "Most galaxies are between 10 billion and 13.6 billion years old."[135] As a consequence of this statement, a minimum of 50% of the galaxies + 1 galaxy for data is expected to fall in the range of 10 billion to 13.6 billion years old. If one were to make the incorrect assumption of a uniform distribution (which would assume and make the median age close to 6.9 billion years) as well as make the assumption that only random chance is at work, the probability of one galaxy falling on Day Seven would be about 2.6%. Likewise, the probability of one galaxy not being created on Day Seven would be 97.4%; because 100% - 2.6% = 97.4%, using the mathematical Law of Total Probability. 97.4% can be represented with the percent sign as the number .974 . The chance of none of the 100 billion galaxies falling on Day Seven would be $(.974)^{100000000000}$ under these assumptions. This is a number too small for most calculators to calculate and they are likely to round the number to zero, but how small this number is can be understood using exponent and logarithm rules along with a little algebra – math comes out to $1 / 10^{1144104312.14}$. So the overall chance of none of the galaxies falling on Day

Seven happening randomly would be less than about $1 / 10^{11000000000}$, or in other words, about 1 over the number 1 followed by more than 1.1 billion zeroes. This would have meant that the overall chance of none the galaxy formations occurring on Day Seven is less than 1 over the number 1 followed a billion zeroes -- if a uniform distribution had been an accurate assumption (it is not). But how far off is it? Since I don't have access to the data and NASA does, I would be very interested in knowing the age distribution and age data for all the galaxies that they've studied.

Atheism is tied to making the argument that both witnesses made their statements and got lucky by chance. Thus, Atheism inherits this number as the chance that the Genesis Writer and St. Peter would both get lucky at implying that on Day Seven, that the number of the heavens were complete and consequently that no new galaxies were formed, but unfortunately due to unavailability of data, I cannot quantify how small this number is.

Chapter 34: The Maximum Probability for Atheism

Atheism with respect to the God of the Scriptures (or any purported deity) as a school of thought must maintain that there are no people with divine knowledge. If there is a single person with divine knowledge, Atheism would be incorrect, since divine knowledge implies the existence of a divine source. The words of the Genesis Writer and the words of Saint Peter with respect to the Genesis Creation Account put Atheism to the test with respect to the Biblical God's existence. Atheism is forced to make the case that neither the Genesis Writer nor St. Peter were divinely inspired. That stance has a near-impossible chance of being correct – as that stance is forced to rely on the Genesis Writer and Saint Peter just getting all the Science correct by chance. The correctness of that atheistic argument is equal to the probability of all the evidence lining-up by chance, including passing the no-new-galaxies-on-day-seven test. In other words, the BEST chance for Atheism's correctness is passing the three tests before it: the Genesis Writer giving the correct relative order of the 13 Day 1 thru Day 6 scientific events I listed is Test One. The placement by the Genesis Writer and by Saint Peter of each of the 13 scientific events with each event falling on the correct time slot is Test Two. The Genesis Writer giving the Day Seven description and Saint Peter giving the Day Seven time slot length with no-new-galaxies-on-Day-Seven is Test Three. To have any chance at correctness, Atheism must argue that passing all three tests occurred by chance. So in this way, the chance of Theism passing these 3 tests randomly is equal to Atheism's best case for survival as a school of thought.

Chance of Atheism's correctness = (pass Test One) x (pass Test Two) x (pass Test Three) = (Chance of passing test of Genesis Writer giving correct order randomly) x (Chance of passing test of the Genesis Writer and Saint Peter giving correct Day Time placement randomly) x (chance of passing the no-new-galaxies-on-day-seven test)

= (1 / 6,227,020,800) x (1 / 793,714,773,254,144) x (chance of passing the no-new-galaxies-on-day-seven-test)

= (1 / 4,942,478,402,320,838,374,195,200) x (chance of passing the no-new-galaxies-on-day-seven-test)

The term in front is approximately 1 / (4.9 x 10^{24}). So

= (1 / (4.9 x 10^{24})) x (chance of passing the no-new-galaxies-on-day-seven-test)

If we ignore the galaxies part, 1 / 4,942,478,402,320,838,374,195,200 is the chance of Atheism being correct under its best-case assumptions. So the chance of Atheism being correct under its best-case assumptions is a less than 1 in 4 Septillion shot. This is the chance of the correctness of Atheism with respect to the non-existence of the Biblical God.

For Atheism to not make the argument that the Genesis Writer and Saint Peter guessed and got lucky by chance, would mean that something other than chance influenced their knowledge base at a time when that was not possible by secular means. This is exactly what Theism would maintain happened: divine inspiration – a supernatural influence gave the Genesis Writer and / or Saint Peter divine knowledge. The Genesis Writer and Saint Peter lived at a time when the knowledge was not available to determine the scientific order and timing of the 13 referenced events nor knowledge through natural means about the times of the formations of galaxies.

Chapter 35: The Minimum Probability for the Existence of God

God exists, or he does not. This gives us the following equation:

Chance God actually exists + Chance God does not actually exist = 1

One of these quantities is zero, the other is 1, which makes the equation true. The Science clearly matched the Bible's testimony. For some, this is enough! It is enough to decide that yes, God in fact does exist!

Others will reach the decision that God exists based on the set of implications regarding the Genesis Writer, which is considered by tradition to be a divinely inspired Moses, or regarding Saint Peter, or both. If either is divinely inspired, you get the existence of God.

But suppose there is a skeptic, held firm in the hands of doubt. This doubting skeptic might say, the best we can do is make estimates to weigh the relative probabilities of each of these. Yet even then, a rational approach would maintain that 'Hey, at the end of the day, he either exists, or he does not. So the equation still needs to equal 1.' Below the skeptic's version of the equation regarding God's existence.

(Man's estimate of the Chance that God exists) + (Man's estimate of the Chance that God does not exist) = 1

This book has shown that the best case for (Man's estimate of the Chance that God does not exist) is that the number is 1 / 4,942,478,402,320,838,374,195,200, which can also be written as approximately $2.02327642 \times 10^{-25}$.

Now for the estimate for the Existence of God.

(Man's estimate of the Chance that God exists) + (Man's estimate of the Chance that God does not exist) = 1

(Man's estimate of the Chance that God exists) + 1 / 4,942,478,402,320,838,374,195,200 = 1

Man's estimate of the Chance that God exists = 1 - 1 / 4,942,478,402,320,838,374,195,200

Man's estimate of the Chance that God exists =
0.9999999999999999999997

Converted to a percentage is to multiply it by 100 and put a percent sign after it.

Man's skeptical estimate of the Chance that God exists =
99.99999999999999999997 %

This is enough information to reach the decision, rationally based on science, that 'Yes, God exists.'

Conclusion: Science Confirms the Existence of God

When the worst case for the Biblical God's existence when weighing probabilities is that God is 99.99999999999999999999997 % likely to exist, there is enough certainty to make a rational and science-based decision that yes, God exists.

This necessitates a dramatic change from the common secular narrative. The common secular narrative's position is that there are significant deep contradictions between the Bible and Science. This narrative is now found to not be real. Instead, a fair, rational, and fact-based estimate equal to or better than 99.99999999999999999999997 % can be generated from which one can discern that God's existence is demonstrated. A significant reason given for discounting the authority of the Scriptures fell apart with the evidence gathered in this book.

Modern Science has confirmed the seven days of creation. This in turn implies that the people, and the records that attested to these scientific facts accurately before natural means were available, were divinely inspired and this also makes the quoted portions of the Book of Genesis and Peter's 2nd Letter divinely inspired. The authorship of the Book of Genesis is ascribed to Moses by tradition[136]. Saint Peter's statement in 2Peter 3:8 provided the timing that was essential to demonstrate, by human observation of scientific facts, the truth of the Seven Days of Creation.

More can be said about Peter.

Peter acknowledged that Jesus was the Messiah, the Son of the living God!

Now when Jesus came into the parts of Caesarea Philippi, he asked his disciples, saying, "Who do men say that the Son of Man is?"

They said, "Some say John the Baptist, some, Elijah, and others, Jeremiah, or one of the prophets."

He said to them, "But who do you say that I am?"

Simon Peter answered, "You are the Messiah, the Son of the living God."

(Gospel of Matthew 16: 13-16, footnotes removed)

And Jesus accepted this acknowledgment and blessed Saint Peter.

Jesus answered him, "Blessed are you, Simon Bar Jonah, for flesh and blood has not revealed this to you, but my Father who is in heaven. I also tell you that you are Peter, and on this rock I will build my church, and the gates of hell will not prevail against it. I will give to you the keys of the Kingdom of Heaven, and whatever you bind on earth will have been bound in heaven; and whatever you release on earth will have been released in heaven."

(Matt 16:13-16, NHEB, footnotes removed)

Jesus stated "Blessed are you, Simon Bar Jonah, for flesh and blood has not revealed this to you, but my Father who is in heaven. I also tell you that you are Peter, and on this rock I will build my church, and the gates of hell will not prevail against it." Jesus spoke truly, because the factual truth of God's existence is the rock upon which his Church depends, which is what we gained from Saint Peter's testimony!

Perhaps it is time to take another look earlier in Saint Peter's 2nd Letter, where he acknowledges that Jesus Christ is the Beloved Son of God and Lord:

For we did not follow cunningly devised tales, when we made known to you the power and coming of our

Lord Jesus Christ, but we were eyewitnesses of his majesty. For he received from God the Father honor and glory, when the voice came to him from the Majestic Glory, "This is my beloved Son, in whom I am well pleased." We heard this voice come out of heaven when we were with him on the holy mountain.

We have the more sure word of prophecy; and you do well that you heed it, as to a lamp shining in a dark place, until the day dawns, and the morning star arises in your hearts: knowing this first, that no prophecy of Scripture is of private interpretation. For no prophecy ever came by the will of man: but men spoke from God, being moved by the Holy Spirit. (2Peter 1:16 – 21, NHEB, footnotes removed)

God exists and He has communicated with us through the Genesis Writer and Saint Peter.

There is a God. There is a source of Hope.

Acknowledgments

Thank you Dr. Gordon Guay for the help and for the discussions.

Thank you Dr. Art West for critiquing an early version of this book and for indicating to me that "I suspect that a lot of Christians won't accept the Book of Jubilees as a divinely inspired source." I also thank Dr. Art West for his ongoing feedback.

Thank you Richard Odom for advising me to use a 365-day calendar year rather than the 364-day based Jubilees Calendar year because the difference would be insignificant. Thank you for advising me to not use the Book of Jubilees as a reference.

Thank you Dr. Bernard Leikind for my increased understanding of the time frame for a day in the Jewish tradition and for pointing out that in *Young's Literal Translation* of the Bible, the evening leads before the day.

Thank you John Kwiatek for the discussions, encouragement, and your editorial support.

Thank you Nancy Jones for your feedback and your skillful editing of Chapter Six.

Thank you Pastor Lynn Samaan, for your encouragement and for telling me about the Blue Letter Bible.

Thank you Michael Ornellas for your encouragement and comments which were constructive in making the manuscript more readable.

Thank you Dr. Jeffrey 'Kapali' Lyon for your replies to my questions about the ancient manuscripts related to the original Greek and Hebrew languages.

I thank my Dad, Eugene Ornellas, for helpful discussions and guidance.

I thank my Mom, Linda Ornellas, for editing and for providing feedback and guidance.

Thank you Malia and the Hill family for encouragement and feedback.

Thank you Pastor Jeff Busby for your encouragement.

Thank you Susan and Mark Doolittle for your help, your encouragement, and your research.

I thank the blueletterbible.org for allowing me to use images from your website in this publication.

Thank you James (Jim) Pacheco for discussions.

Blessings for the memory of my deceased Godmother, Margaret T. O'Connor.

Appendix A: Julian Calendar versus the Gregorian Calendar

Suppose an interested party asked the question "How many days are there in a year?" If we are talking about the Julian Calendar, in a non-leap year of the Julian Calendar, there would be 365 days in a year. If we are talking specifically about a leap year within the Julian Calendar, the answer would be 366 days in a year. A leap year in the Julian Calendar occurs once every 4 years. Since the leap year occurs once every 4 years, calculating the average length of the year in the Julian Calendar goes as follows:

$$(3 \times 365 \text{ days} + 1 \times 366 \text{ days}) / 4 \text{ years} = 365.25 \text{ days per year.}$$

This yields a different length in Julian years than the 365.0 days I used throughout the rest of this book, making the rigorous calculation for Saint Peter's statement (when including leap years) to the following:
1 Day with the Lord =1000 years Heaven Time x (365.25 Days Heaven Time / Years Heaven Time) x 1000 Julian years of our time / 1 Day of Heaven Time = 365,250,000 Julian years.
This in turn gives a different timetable for Genesis:

TABLE 24: The Timetable of Genesis using the Ratio that Includes the Leap Year's Contribution for the Julian Calendar

Timeline in Genesis 1 & 2	**Nights and Days that remain until the end of Day 7**	**Units of the Lord's Time that Remain until the End of Day 7.**	**Calculated Timeline Based on 2Peter 3:8 (years ago)**
Beginning of Night One	7 nights + 7 days	14	5,113,500,000 (5.1135 billion years ago)
Beginning of Day One	6 nights + 7 days	13	4,748,250,000
Beginning of Night Two	6 nights + 6 days	12	4,383,000,000
Beginning of Day Two	5 nights + 6 days	11	4,017,750,000

Beginning of Night Three	5 nights + 5 days	10	3,652,500,000
Beginning of Day Three	4 nights + 5 days	9	3,287,250,000
Beginning of Night Four	4 nights + 4 days	8	2,922,000,000
Beginning of Day Four	3 nights + 4 days	7	2,556,750,000
Beginning of Night Five	3 nights + 3 days	6	2,191,500,000
Beginning of Day Five	2 nights + 3 days	5	1,826,250,000
Beginning of Night Six	2 nights + 2 days	4	1,461,000,000
Beginning of Day Six	1 night + 2 days	3	1,095,750,000
Beginning of Night Seven	1 night + 1 day	2	730,500,000
Beginning of Day Seven	1 day	1	365,250,000
End of Day Seven	None	0	0 (6,000 to 10,000 years ago exit from Eden timespan is discarded)
The fourth column is the result of multiplying the third column by 365,250,000 years			

While there are exactly 365.25 days per year on average in the Julian Calendar, there are approximately 365.2425 days per year on average in the modern Gregorian Calendar. That is, the calendar systems average length for a year differ only by about .0075 days, which is less than .0021 % of the length of a modern Georgian Calendar year. The tiny difference between the two calendar systems means that the modern Gregorian Calendar year is on average about 10 minutes and 48 seconds shorter than the average length of a Julian Calendar year. Moreover, since the Julian Calendar is so much easier to work with to provide precise times, this is the preferred calendar of modern Astronomy as well.

Suppose someone questioned if the Julian calendar system was accurate and wanted to use the Gregorian calendar instead[137] or if that person wondered whether perhaps St. Peter was talking in absolute time units, would that introduce enough deviation into the above calculation to possibly change the conclusions? The answer appears to be 'No'; it would not introduce enough deviation into the above calculation to change the conclusions. According to a website authored by Dirk Husfeld, the average length of a Julian calendar year is 365.25 days UT per year, and the average length of a Gregorian calendar year is 365.2425 days UT.[138] UT is Universal Time, which is a time standard that involves the average speed of the Earth's rotation.[139]

The calculation runs as follows: ignore the 10 days being dropped in October 1582, (I'm already throwing out modern human history).[140] Next, I subtract the average length of the Gregorian calendar year from the average length of a Julian calendar year. This gives me the difference in time in days per year advantage of Julian over Gregorian, which is .0075 days per year. Next, I want to see the max deviation in Julian years (so I can compare it to my other calculations), so I divide this by the days to year ratio of a Julian calendar year to get the average length difference in Julian calendar years that the Gregorian calendar would imply. I then multiply by the amount of time implied by St. Peter's 2 Peter 3:8 statement. This gives the following result:

365.25 – 365.2425 = .0075 days

.0075 days x (1 Julian year / 365.25 days) = 2.05338809E-5 Julian years

Approximation of Max deviation = 5,113,500,000 x 2.05338809E-5 = 105,000 Julian years.

Thus, this is the approximate amount a Gregorian calendar-bases calculation would deviate from the Julian calendar for the bulk time. Approximately 105,000 Julian calendar years does not appear to significantly alter the outcomes involved in the calculations in the body of the book.

If we had wanted to assume that Saint Peter was using our modern calendar system or something close to it, we could have approximated this quantity using the Gregorian Calendar, which has an average length in days per year of approximately 365.2425 days per year[141]. The resulting approximation is shown below.

1 Day of the Lord's time would approximately equal 365,242,500 years of Gregorian Calendar time.

That is approximately 1 Day of the Lord's time = 365,242,500 years of our (modern) time.

Throughout this book, I used the 1 day of the Lord's time = 365,000,000 Julian years for the reasons I discussed in Chapter Ten – to ease the burden of calculating the calculation for readers who are not mathematicians – even though this comes with a small error.

Appendix B: Supplementary Figures

A collection of supplementary figures is below.

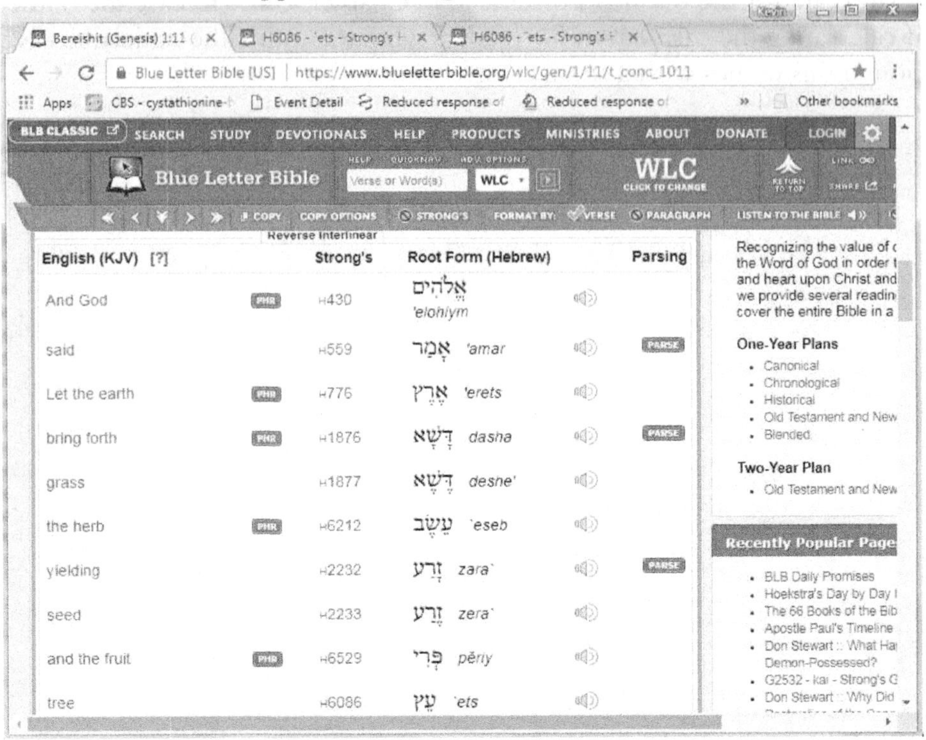

Figure 1 (Bereishit (Genesis) 1:11 (WLC) -.)

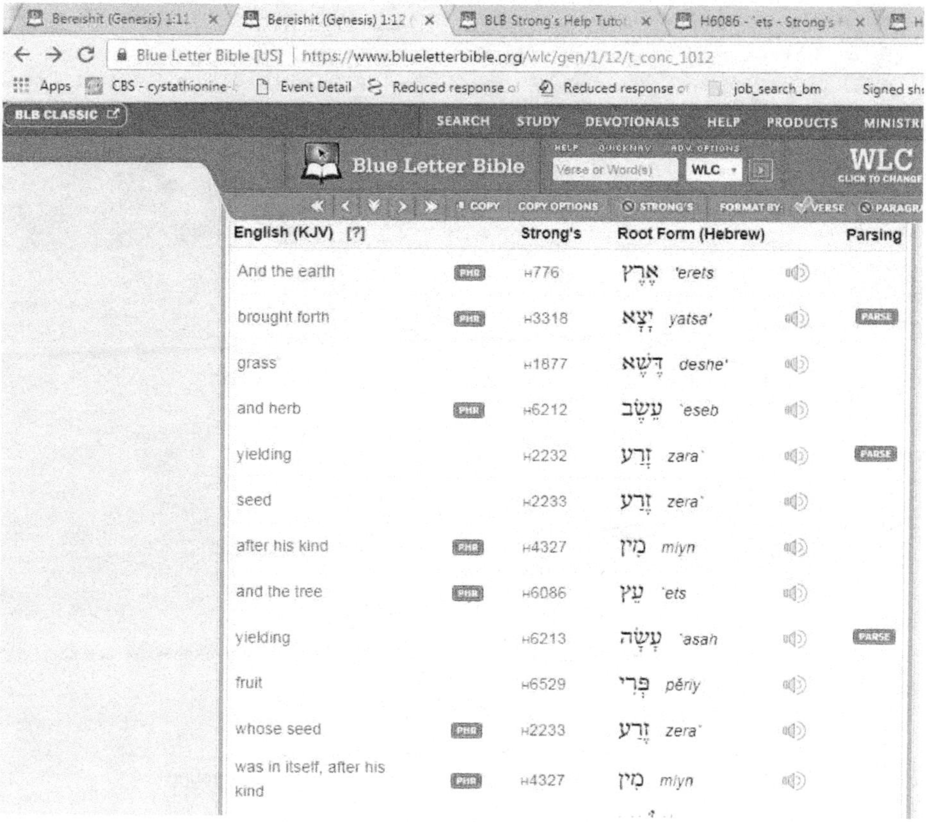

Figure 2 (Bereishit (Genesis) 1:12 (WLC) -.)

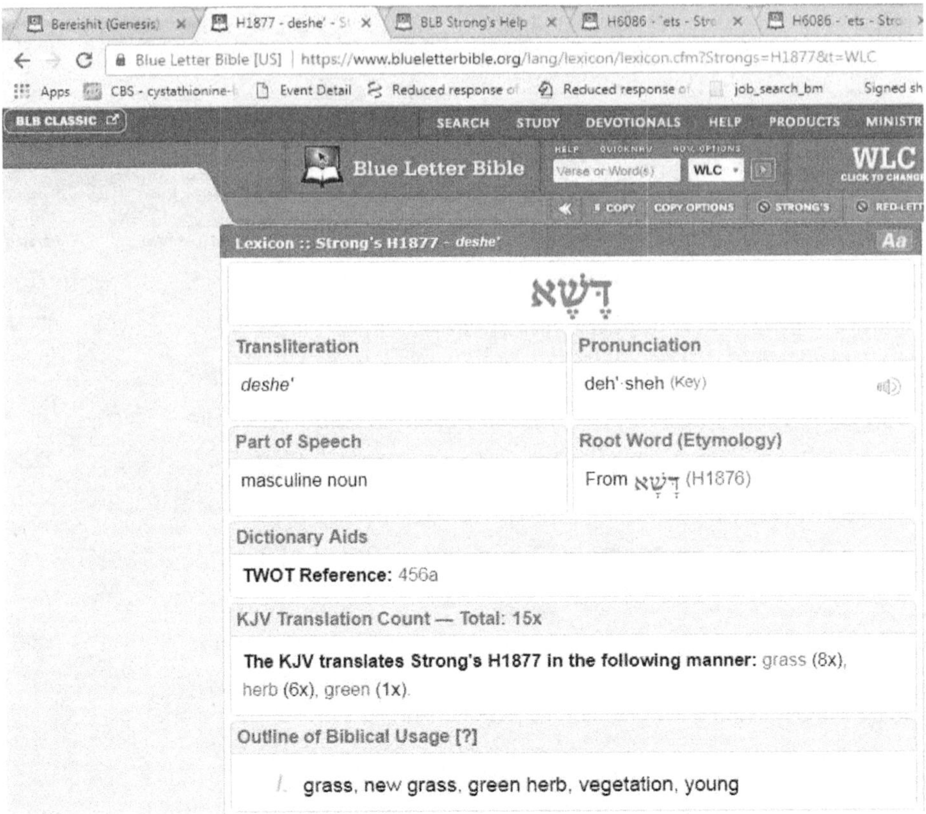

Figure 3 deshe' [142]

Figure 4 Strong's Definitions of deshe' showing green in the list[142]

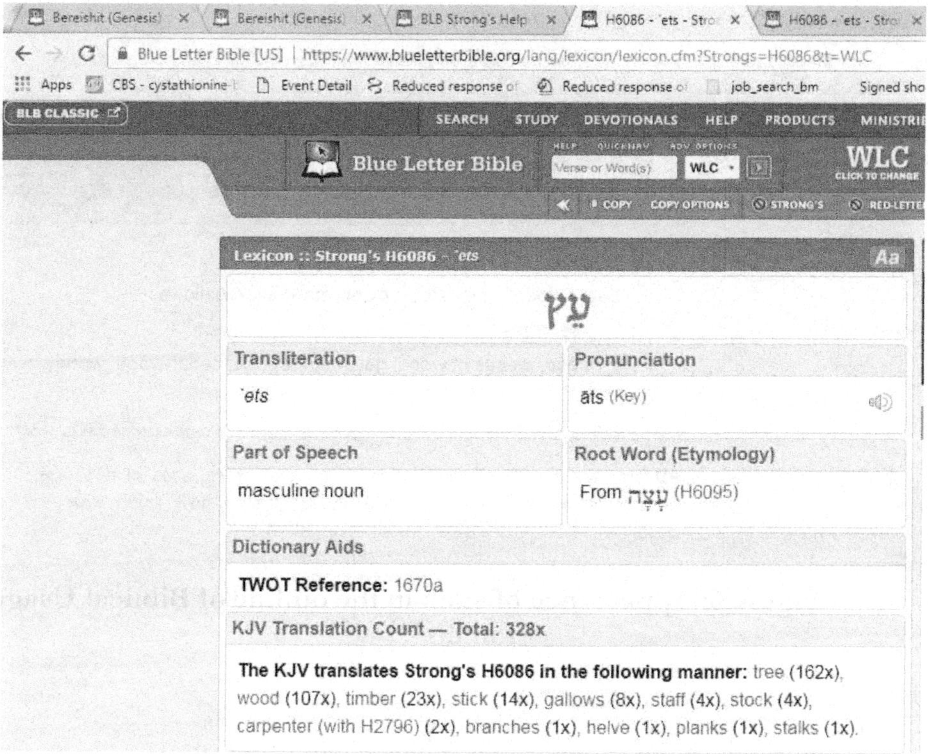

Figure 5 Strong's 'ets (H6086 - `ets - Strong's Hebrew Lexicon (WLC).)

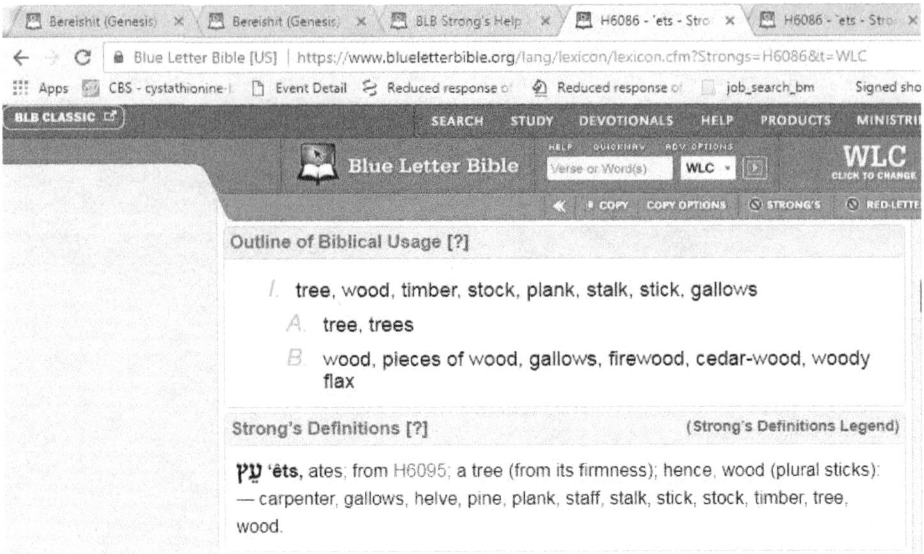

Figure 6: Appearance of stalk in the outline of Biblical Usage for 'ets [143]

Appendix C: The Number of Galaxies

What is the number of galaxies? From the official NASA Tumblr account:

"We already estimated that there were about 100 billion galaxies in the observable universe, but new research shows that this estimate is at least 10 times too low!

First, what is the observable universe? Well, it is the most distant part of the universe we can see from Earth because, in theory, the light from these objects have had time to reach Earth.

In a new study using surveys taken by the Hubble Space Telescope and other observatories, astronomers came to the surprising conclusion that there are at least 10 times more galaxies in the observable universe than previously thought. This places the universe's estimated population at, minimally, 2 trillion galaxies!
...
Using deep-space images from the Hubble Space Telescope and other observatories, astronomers converted the images into 3-D, in order to make accurate measurements of the number of galaxies at different epochs in the universe's history.

In addition, they used new mathematical models, which allowed them to infer the existence of galaxies that the current generation of telescopes cannot observe. This led to the surprising conclusion that in order for the numbers of galaxies we now see and their masses to add up, there must be a further 90% of galaxies in the observable universe that are too faint and too far away to be seen with present-day telescopes.
The myriad small faint galaxies from the early universe merged over time into the larger galaxies we can now observe.

That means that over 90% of the galaxies in the universe have yet to be studied! In the near future, the James Webb Space Telescope will be able to study these ultra-faint galaxies and give us more information about their existence."[144]

This 2016 post links to https://www.nasa.gov/feature/goddard/2016/hubble-reveals-observable-universe-contains-10-times-more-galaxies-than-previously-thought when the '10 times more galaxies' link is clicked.

The estimated number of galaxies now is at least 2 trillion galaxies.

References

(auto-generated), c. (n.d.). *Inclusions - Termwiki, millions of terms defined by people like you.* Retrieved July 25, 2017, from termwiki: https://ue.termwiki.com/UE/inclusions_%E2%82%84

29.2B: Gnathostomes - Jawed Fishes - Biology LibreTexts. (2021, March 6). Retrieved September 26, 2021, from bio.libretexts.org: https://bio.libretexts.org/Bookshelves/Introductory_and_General_Biology/Book%3A_General_Biology_(Boundless)/29%3A_Vertebrates/29.2%3A_Fishes/29.2B%3A_Gnathostomes_-_Jawed_Fishes

Amelin, Y., Kaltenbach, A., Iizuka, T., Stirling, C. H., Ireland, T. R., Petaev, M., & Jacobsen, S. B. (2010). U–Pb chronology of the Solar System's oldest solids with variable 238U/235U. *Earth and Planetary Science Letters, 300*(3-4), 343-350. doi:10.1016/j.epsl.2010.10.015

Barton, N. H., Briggs, D. E., & Eisen, J. A. (2007). *Evolution.* New York, United States of America: Cold Spring harbor Laboratory Press.

Battistuzzi, F. U., Feijao, A., & Hedges, S. B. (2004, November 9). A genomic timescale of prokaryote evolution: insights into the origin of methanogenesis, phototrophy, and the colonization of land. *BMC Evolutionary Biology volume, 4*(44). doi:10.1186/1471-2148-4-44

Bereishit (Genesis) 1:11 (WLC) -. (n.d.). Retrieved August 3, 2017, from Blue Letter Bible: https://www.blueletterbible.org/wlc/gen/1/11/t_conc_1011

Bereishit (Genesis) 1:12 (WLC) -. (n.d.). Retrieved August 3, 2017, from Blue Letter Bible: https://www.blueletterbible.org/wlc/gen/1/12/t_conc_1012

Betts, H. C., Puttick, M. N., Clark, J. W., Williams, T. A., Donoghue, P. C., & Pisani, D. (2018, October). Integrated genomic and fossil evidence illuminates life's early evolution and eukaryote origins. *Nature Ecology & Evolution, 2*(10), 1556-1562. doi:10.1038/s41559-018-0644-x

Bikos, K., & Jones, G. (n.d.). *What is Universal Time (UT)?* Retrieved from timeanddate: https://www.timeanddate.com/time/universal-time.html

Bond, C., & Harris, A. (1988, June). Locomotion of sponges and its physical mechanism. *The Journal of Experimental Zoology, 246*(3), 271-284. doi:10.1002/jez.1402460307

Boulton, G. (1987). PROGRESS IN GLACIAL GEOLOGY DURING THE LAST FIFTY YEARS. *Journal of Glaciology*, 25-32. Retrieved September 12, 2021, from https://www.researchgate.net/profile/Geoffrey-Boulton/publication/323419049_Progress_in_glacial_geology_during_the_last_fifty_years/links/5a992a67aca27214056e0fec/Progress-in-glacial-geology-during-the-last-fifty-years.pdf

Bouvier, A., & Wadhwa, M. (2010, August). The age of the Solar System redefined by the oldest Pb-Pb age of a meteoritic inclusion. *Nature Geoscience*, 1-5. doi:10.1038/ngeo941

Britannica, The Editors of Encyclopaedia. (2017, June 1). *Craton*. Retrieved February 6, 2021, from Encyclopedia Britannica: https://www.britannica.com/science/craton

Burley, S. (2013). Time and motion. In M. Hall (Ed.), *52 Things about Geology* (pp. 100-101). Agile Libre. Retrieved from https://www.researchgate.net/profile/Stuart-Burley/publication/323969982_Time_and_motion/links/5ab5549f0f7e9b68ef4cf1d3/Time-and-motion.pdf

Cain, F. (2004, December 1). *Youngest Galaxy Found - Universe Today*. Retrieved December 19, 2021, from universetoday.com: https://www.universetoday.com/10075/youngest-galaxy-found/

Calvin, M. (1965). *Chemical Evolution*. United States. doi:10.2172/919675

Cattermole, P., & Moore, P. (1985). *The Story of Earth*. Cambridge, United Kingdom: Cambridge University Press. Retrieved October 2020, from https://www.google.com/books/edition/The_Story_of_the_Earth/1xs9AAAAIAAJ?hl=en&gbpv=1

Cleeves, I. L., Begin, E. A., Alexander, C. M., Du, F., Graninger, D., Oberg, K. I., & Harries, T. J. (2014, September 25). *The Ancient Heritage of Water Ice in the Solar System*. Retrieved from arxiv.org: https://arxiv.org/pdf/1409.7398.pdf

Cohen, K., Finney, S., Gibbard, P., & Fan, J. -X. (2013; updated (requested citation of work)). ChronostratChart2017-02.jpg (4794×3368). *The ICS International Chronostratigraphic Chart(36)*, 199-204. International Commission on Stratigraphy. Retrieved September 26, 2020, from https://stratigraphy.org/ICSchart/ChronostratChart2017-02.pdf

Cornell, D. H. (1993). Rare earths from supernova to superconductor. *Pure and Applied Chemistry, 65*(12), 2453-2464. Retrieved from https://www.degruyter.com/downloadpdf/journals/pac/65/12/article-p2453.pdf

Dalrymple, B. G. (2004). *Ancient Earth, Ancient Skies: The Age of Earth and Its Cosmic Surroundings.* Stanford University Press.

Diaz, J. M., & Laybourn-Parry, J. E. (2019, February 7). *protozoan.* Retrieved June 3, 2021, from Encyclopedia Britannica: https://www.britannica.com/science/protozoan

Dunford, B. (2020, September 16). *Planets Sun: In Depth.* Retrieved September 26, 2020, from Solar System Exploration NASA Science: https://solarsystem.nasa.gov/planets/sun/indepth

Eicher, D. J. (2020, May 19). *How Many Galaxies Are There? Astronomers Are Revealing the Enormity of the Universe: The universe is awash in islands of matter — some 100 billion galaxies make up the basic building blocks of the cosmos.* Retrieved December 19, 2021, from discovermagazine.com: https://www.discovermagazine.com/the-sciences/how-many-galaxies-are-there-astronomers-are-revealing-the-enormity-of-the

Eldredge, N. (Ed.). (2002). *Life on Earth: An encyclopedia of Biodiversity, Ecology, and Evolution* (Vols. 1 A-G). Santa Barbara, California. Retrieved from https://books.google.com/books?id=iVEWPg8vnxgC&printsec=frontcover&source=gbs_ge_summary_r&cad=0#v=onepage&q&f=false

Encyclopædia Britannica, Inc. (2021). *Hydrosphere - origin and evolution.* Retrieved January 28, 2021, from britannica.com: https://www.britannica.com/science/hydrosphere/Origin-and-evolution-of-the-hydrosphere

Engelking, C. (2014, September 25). *Earth's Water Is Older Than The Sun | Discover Magazine.* Retrieved from discovermagazine.com:

https://www.discovermagazine.com/the-sciences/earths-water-is-older-than-the-sun

Eriksson, P. G., Mazumder, R., Catuneanu, O., Bumby, A. J., & Ilondo, B. O. (2006). Precambrian continental freeboard and geological evolution: A time perspective. *Earth-Science Reviews, 79*, 165-204. doi:10.1016/j.earscirev.2006.07.001

Fleur, N. S. (2016, April 15). The Water in Your Glass Might Be Older Than the Sun. www.nytimes.com. Retrieved February 21, 2018, from https://www.nytimes.com/2016/04/16/science/the-water-in-your-glass-might-be-older-than-the-sun.html

Forbes, D. A., Ponman, T. J., & Brown, R. J. (1998, October 15). Dependence of the Fundamental Plane Scatter on Galaxy Age. *The Astrophysical Journal, 508*(1), L43-L46. doi:10.1086/311715

Freeman, S. (2005). Biological Science. In S. Freeman, *Biological Science* (2nd ed.). Upper Saddle River, NJ, USA: Pearson Education, Inc.

Galperin, M. Y. (2013). Genome Diversity of Spore-Forming Firmicutes. *Microbiology Spectrum*, 1-15.

Gen 1 | ESV | STEP | In the beginning, God created the heavens and the ... (2021). Retrieved September 23, 2021, from stepbible.org: www.stepbible.org

Genitsaris, S., Moustaka-Gouni, M., & Kormas, K. A. (2011, January 19). Airborne microeukaryote colonists in experimental water containers: diversity, succession, life histories and established food webs. *Aquatic Microbial Ecology, 62*, 139-152. doi:10.3354/ame01463

Gopalan, K., Kaushal, S. L.-H., & Wetherill, G. (1970). Rubidium-Strontium, Uranium, and Thorium-Lead Dating of Lunar Material. *Science, 167*(3918), 471-473. doi:10.1126/science.167.3918.471

Goudie, A., & Cuff, D. J. (2002). *Encyclopedia of Global Change: J-Z.* United Kingdom: Oxford University Press. Retrieved from https://books.google.com/books?id=5YrqQqW11aYC&q=earth#v=snippet&q=earth%20history&f=false

H1877 - deshe' - Strong's Hebrew Lexicon (WLC). (n.d.). Retrieved August 3, 2017, from Blue Letter Bible:

https://www.blueletterbible.org//lang/lexicon/lexicon.cfm?Strongs=H1877&t=WLC

H3671 - kānāp̄ - Strong's Hebrew Lexicon (kjv). (n.d.). Retrieved May 11, 2022, from blueletterbible.org: https://www.blueletterbible.org/lexicon/h3671/kjv/wlc/0-1/

H6086 - `ets - Strong's Hebrew Lexicon (WLC). (n.d.). Retrieved August 3, 2017, from Blue Letter Bible: https://www.blueletterbible.org/lang/lexicon/lexicon.cfm?Strongs=H6086&t=WLC

Hays, J. F., & Walker, D. (1977). Lunar Igneous Rocks and the Nature of the Lunar Interior. *The Soviet-American Conference on Cosmochemistry of the Moon and Planets, Part 1* (pp. 127-135). Washington: United States. Retrieved from https://ntrs.nasa.gov/archive/nasa/casi.ntrs.nasa.gov/19780004989.pdf

Hocken, V. (n.d.). *The Gregorian Calendar.* Retrieved from timeanddate.com: https://www.timeanddate.com/calendar/gregorian-calendar.html

How Old Are Galaxies? | NASA Space Place – NASA Science for Kids. (2019, January 14). Retrieved from Space Place Explore Earth and Space at NASA.gov: https://spaceplace.nasa.gov/galaxies-age/en/

Husfeld, D. (1997, July 17). *Astronomical Time Keeping.* (C. Kronberg, Editor) Retrieved September 28, 2020, from The Munich Astro Archive: http://www.maa.clell.de/Scholar/calendar.html

Huynh, M. (2019, March 5). *Clues to the Early Rise of Oxygen on Earth Found in Sedimentary Rock.* Retrieved from NASA Astrobiology Institute: https://astrobiology.nasa.gov/nai/articles/2019/3/5/clues-of-earths-early-rise-of-oxygen/index.html

Knoll, A., Javaux, E., Hewitt, D., & Cohen, P. (2006, May 5). Eukaryotic organisms in Proterozoic oceans. *Philosophical Transactions of the Royal Soceity B, 361*, 1023-1038. doi:10.1098/rstb.2006.1843

Kumar, S., Stecher, G., Suleski, M., & Hedges, S. B. (2017). TimeTree: a resource for timelines, timetrees, and divergence times. *Molecular Biology and Evolution, 34*, 1812-1819. doi:10.1093/molbev/msx116

Liu, H., Zartman, R. E., Ireland, T. R., & Sun, W.-d. (2019, September). Global atmospheric oxygen variations recorded by Th/U systematics of igneous rocks. (M. H. Thiemens, Ed.) *Proceedings of the National Academy of Sciences of the United States of America, 116*(38), 18854-18859. doi:10.1073/pnas.1902833116

Lowman Jr., Paul D (NASA Goddard Space Flight Center Greenbelt, MD, United States). (1971). *THE GEOLOGIC EVOLUTION OF THE MOON (Revised Version).* NASA, Goddard Space Flight Center. Greenbelt, Maryland: Journal of Geology. Retrieved from https://ntrs.nasa.gov/archive/nasa/casi.ntrs.nasa.gov/19730014050.pdf

Mayer, W. (Ed.). (2014). International Commission on the History of Geological Sciences INHIGEO ANNUAL RECORD No. 46 Covering activities generally in 2013. 56-58. (W. Mayer, Compiler) Canberra. Retrieved October 6, 2020, from https://shareok.org/bitstream/handle/11244/44844/INHIGEO-2014-46.pdf?sequence=7

Moore, V. (1993, June 16). *JOSEPH SAUNDERS DIES AT 66 - The Washington Post.* Retrieved from The Washington Post: https://www.washingtonpost.com/archive/local/1993/06/16/joseph-saunders-dies-at-66/4b6f487b-506d-4fff-b56e-3ae28829cec3/

Mosley Jr., J. (2021, July). *Who Wrote Genesis? - Heidelberg Seminary.* Retrieved from Heidelberg Theological Seminary: https://heidelbergseminary.org/2021/07/who-wrote-genesis/

Mukhopadhyay, J., Crowley, Q. G., Ghosh, S., Chakrabarti, K., Misra, B., Heron, K., & Bose, S. (2015, April). Forum Reply: Oxygenation of the Archean atmosphere: New paleosol constraints from eastern India. *Geology, 42*(10), 923-926. doi:10.1130/G36091.1

NASA. (2016, October 13). *NASA --- 10 Times More Galaxies!* Retrieved from Tumblr.com: https://nasa.tumblr.com/post/151753781974/10-times-more-galaxies?linkId=29991157

NASA. (2019, January 24). *How Old Are Galaxies? | NASA Space Place -- NASA Science for Kids.* Retrieved from NASA Science - SpacePlace: https://spaceplace.nasa.gov/galaxies-age/en/

NASA Glenn Research Center. (n.d.). *Calendar Calculations*. Retrieved from nasa.gov: https://www.grc.nasa.gov/www/k-12/Numbers/Math/Mathematical_Thinking/calendar_calculations.htm

NASA Official: Dr. Edward J. Wollack. (n.d.). *WMAP - Age of the Universe*. Retrieved from National Aeronautics and Space Administration: https://wmap.gsfc.nasa.gov/universe/uni_age.html

National Aeronautics and Space Administration. (1971). *The Experiments of Biosatellite II*. (J. F. Saunders, Ed.) Washington, D.C.: Scientific and Technical Information Office: National Aeronautics and Space Administration.

Nichols, S., & Wörheide, G. (2005, April 1). Sponges: New Views of Old Animals. *Integrative and Comparative Biology, 45*(2), 333-334. doi:https://doi.org/10.1093/icb/45.2.333

Noffke, N., Christian, D., Wacey, D., & Hazen, R. M. (2013). Microbially Induced Sedimentary Structures Recording an Ancient Ecosystem in theca.3.48 Billion-Year-Old Dresser Formation, Pilbara, Western Australia. *Astrobiology, 13*(12), 1103-1124. doi:10.1089/ast.2013.1030

Nutman, A. P., Mojzsis, S. J., & Friend, C. R. (1997). Recognition of ≥3850 Ma water-lain sediments in West Greenland and their significance for the early Archaean Earth. *Geochimica et Cosmochimica, 61*(12), 2475-2484. doi:10.1016/S0016-7037(97)00097-5

Ollier, C. D. (1996). Planet Earth. In I. Douglas, R. Huggett, & M. Robinson (Eds.), *Companion Encyclopedia of Geography: The Environment and Humankind* (pp. 15-43). London: Routledge. Retrieved from https://books.google.com/books?id=8u-JAgAAQBAJ&printsec=frontcover#v=onepage&q&f=false

Papanastassiou, D., & Wasserburg, G. (1971, July 19). Rb-Sr AGES OF IGNEOUS ROCKS FROM THE APOLLO 14 MISSION AND THE AGE OF THE FRA MAURO FORMATION. *EARTH AND PLANETARY SCIENCE LETTERS, 12*, 36-48. Retrieved from https://www.researchgate.net/profile/G-Wasserburg/publication/254258543_Rb-Sr_AGES_OF_IGNEOUS_ROCKS_FROM_THE_APOLLO_14_MISSION_AND_THE_AGE_OF_THE_FRA_MAURO_FORMA

TION_DAPAPANASTASSIOU_and_GJWASSERBURG/links/00b7d51fc55f9ee819000000/Rb-Sr-AGES-OF-IGNEOUS-RO

Peterson, K. J. (1994). The Origin and Early Evolution of the Craniata. *Short Courses in Paleontology, Volume 7: Major Features of Vertebrate Evolution*, 14-37. doi:10.1017/S2475263000001240

Rampazzo, R., Marino, A., Tantalo, R., Bettoni, D., Buson, L. M., Chiosi, C., . . . Rich, R. M. (2007, September 17). The Galaxy Evolution Explorer UV emission in shell galaxies: tracing galaxy 'rejuvenation' episodes. *Monthly Notices of the Royal Astronomical Society, 381*(1), 245–262. doi:10.1111/j.1365-2966.2007.12246.x

Ridley, W. I. (1975). On high-alumina mare basalts. *Proceedings. Lunar Science Conference, 6th. 1*, pp. 131-145. Houston: Pergamon Press, Inc. Retrieved September 13, 2018, from http://adsbit.harvard.edu//full/1975LPSC....6..131R/0000131.000.html

Schlager, W. (2008). *Austrian Journal of Earth Sciences* (Vol. 101). Vienna: Österreichische Geologische Gesellschaft. Retrieved from https://www.zobodat.at/pdf/MittGeolGes_101_0004-0016.pdf

Schlichting Jr., H. E. (1969). The Importance Of Airborne Algae and Protozoa. *Journal of the Air Pollution Control Association, 19*(12), 946-951. doi:10.1080/00022470.1969.10469362

Schweitzer, F., & Seitzer, P. (1998, November). AGES AND METALLICITIES OF YOUNG GLOBULAR CLUSTERS IN THE MERGER REMNANT NGC 7252. *The Astronomical Journal, 116*(5), 2206-2219. doi:10.1086/300616

Seilacher, A., Bose, P. K., & Pflu¨ger, F. (1998, October 2). Triploblastic Animals More Than 1 Billion Years Ago: Trace Fossil Evidence from India. *Science, 282*(80), 80,82. doi:10.1126/science.282.5386.80

Silver, L. T. (1970). Uranium-Thorium-Lead Isotope Relations in Lunar Materials. *Science, 167*(3918), 468-471. doi:10.1126/science.167.3918.468

Sinha, K. A., & Tilton, G. R. (1973). Isotopic evolution of common lead. *Geochimica et Cosmochimica Acta, 37*, 1823-1849. Retrieved from www.researchgate.net/profile/A-Sinha-

3/publication/36097641_The_isotopic_evolution_of_common_lead_and_its_bearing_on_the_age_of_the_earth/links/548f47710cf225bf66a7ff06/The-isotopic-evolution-of-common-lead-and-its-bearing-on-the-age-of-the-earth.pdf

Smith, P. E. (1973). The Effects of Some Air Pollutants and Meteorological Conditions on Airborne Algae and Protozoa. *Journal of the Air Pollution Control Association, 23*(10), 876-880. doi:10.1080/00022470.1973.10469858

Spielman, S. J., & Moore, E. K. (2020, September 24). dragon: A New Tool for Exploring Redox Evolution Preserved in the Mineral Record. *Frontiers in Earth Science, 8*. doi:10.3389/feart.2020.585087/full

Spores | plantbot. (n.d.). Retrieved February 2022, 21, from Plantbot Genetics: https://www.plantbotgenetics.com/spores

Su, D., Yang, L., Shi, X., Ma, X., Zhou, X., Hedges, S. B., & Zhong, B. (2021). Large-Scale Phylogenomic Analyses Reveal the Monophyly of Bryophytes and Neoproterozoic Origin of Land Plants. (F. U. Battistuzzi, Ed.) *Molecular Biology and Evolution.* doi:10.1093/molbev/msab106

Tang, J., & Joyce, M. (2021, May). Revised Best Estimates for the Age and Mass of the Methuselah Star HD 140283 Using MESA and Interferometry and Implications for 1D Convection. *Research Notes AAS, 5*(117). Retrieved from https://iopscience.iop.org/article/10.3847/2515-5172/ac01ca

Tatsumoto, M., & Rosholt, J. N. (1970, January 30). Age of the Moon: An Isotopic Study of Uranium-Thorium-Lead Systematics of Lunar Samples. *Science, 167*(3918), 461-463. doi:10.1126/science.167.3918.461

Tatsumoto, M., Hedge, C., Doe, B., & Unruh, D. (1972). U-Th-Pb and Rb-Sr measurements on some Apollo 14 lunar samples*. *Proceedings of the Third Lunar Science Conference (Supplement 3, Geochimica et Cosmochimica Acta). 2*, pp. 1531-1555. The M.I.T. Press. Retrieved September 12, 2021, from http://articles.adsabs.harvard.edu/cgi-bin/nph-iarticle_query?bibcode=1972LPSC....3.1531T&db_key=AST&page_ind=0&data_type=GIF&type=SCREEN_VIEW&classic=YES

Tatsumoto, M., Nunes, P. D., & Unruh, D. M. (1977). Early History of the Moon: Implications of U-Th-Pb and Rb-Sr Systematics. In J. H. Pomeroy, & N. J. Hubbard (Ed.), *The Soviet-American Conference on Cosmochemistry of the Moon and Planets. 370, Part 2*, pp. 507-523. Scientific and Technical Information Office, National Aeronautics and Space Administration. Retrieved September 12, 2021, from https://books.google.com/books?id=dCkgAAAAIAAJ&printsec=frontcover#v=onepage&q&f=false

Terlevich, A., & Forbes, D. A. (2002, March 1). A catalogue and analysis of local galaxy ages and metallicities. *Monthly Notices of the Royal Astronomical Society, 330*(3), 547–562. doi:10.1046/j.1365-8711.2002.05073.x

The Editors of Encyclopaedia Britannica. (n.d.). *Prokaryote | Definition, Example, & Facts | Britannica.* Retrieved from Encyclopaedia Britannica: https://www.britannica.com/science/prokaryote

The Messianic Jewish Family Bible Society. (2015). *Genesis 1:20-23 TLV - Then God said, "Let the waters swarm - Bible Gateway.* Retrieved May 2020, from Bible Gateway: https://www.biblegateway.com/passage/?search=Genesis+1%3A20-23&version=TLV

Thomassot, E., O'Neil, J., Francis, D., Cartigny, P., & Wing, B. A. (2015, January 20). Atmospheric record in the Hadean Eon from multiple sulfur isotope measurements in Nuvvuagittuq Greenstone Belt (Nunavik, Quebec). (M. H. Thiemens, Ed.) *Proceedings of the National Academy of Sciences, 112*(3), 707-712. doi:10.1073/pnas.1419681112

University of Nottingham. (2012, February 27). *Immortal worms defy aging.* Retrieved March 28, 2022, from ScienceDaily: https://www.sciencedaily.com/releases/2012/02/120227152612.htm

University of Wisconsin - La Crosse: Gillis, Rick; Haro, Roger J. . (2021). *Lab 3: Protozoans - Zoo-lab | UW-La Crosse.* Retrieved from University of Wisconsin - La Crosse Biology Department: https://www.uwlax.edu/biology/zoo-lab/lab-3--protozoans/

van den Heuval, E. (2016). From Universe to Multiverse. *The Amazing Unity of the Universe: And Its Origin in the Big Bang*(2). Springer, Cham. doi:10.1007/978-3-319-23543-1_16

Voelz, H., & Reichenbach, H. (1969, September). Fine Structure of Fruiting Bodies of Stigmatella Aurantiaca (Myxobacterales). *Journal of Bacteriology, 99*(3), 856-866.

Wall, M. (2014, September 25). *Much of Earth's Water Is Older Than the Sun | Space.* Retrieved from space.com: https://www.space.com/27256-earth-water-older-than-sun.html

Williams, G. E. (2000, February). Geological Constraints on the Precambrian History of Earth's Rotation and the Moon's Orbit. *Review of Geophysics, 38*(1), 37-59. doi:10.1029/1999RG900016

Williams, G. E., Schmidt, P. W., & Young, G. M. (2016). Strongly seasonal Proterozoic glacial climate in low palaeolatitudes: Radically different climate system on the pre-Ediacaran Earth. *Geoscience Frontiers*(7), 555-571. doi:10.1016/j.gsf.2016.01.005

www.researchmedia.eu. (2014). *Investigating plate tectonics.* The University of Adelaide, Adelaide. Retrieved October 9, 2020, from https://www.adelaide.edu.au/directory/alan.collins?dsn=directory.file;field=data;id=29772;m=view

Yoshizaki, T., Nakashima, D., Nakamura, T., Park, C., Sakamoto, N., Ishida, H., & Itoh, S. (2019, February 22). *1812.00374.pdf.* doi:10.1016/j.gca.2019.02.034

Endnotes

[3] Part of the verse in Genesis 1:5 from Young's Literal Translation (YLT).
[4] See Table 5 for a list of scientists with dates supporting this specific timing.
[5] (Cleeves, et al., 2014)
[6] (Bouvier & Wadhwa, 2010, p. 1)
[7] (Thomassot, O'Neil, Francis, Cartigny, & Wing, 2015, p. 707) The quote is in the Significance section and within that section's last sentence.
[8] (Cohen, Finney, Gibbard, & Fan, 2013; updated (requested citation of work)) 2017-02 Chart displays. Requested citation to be 2013; updated. It is unclear whether the boundary for the end of the Hadean Eon is 4.0 billion years ago or 4.000 billion years ago. I chose the more conservative, less exacting number for the statement.
[9] (Ollier, 1996, p. 36)
[10] (Nutman, Mojzsis, & Friend, 1997, p. 2475)
[11] (Encyclopædia Britannica, Inc., 2021)
[12] (Eriksson, Mazumder, Catuneanu, Bumby, & Ilondo, 2006, pp. 169,182,192)
[13] (Kumar, Stecher, Suleski, & Hedges, 2017)
[14] (Kumar, Stecher, Suleski, & Hedges, 2017)
[15] (Williams, Schmidt, & Young, Strongly seasonal Proterozoic glacial climate in low palaeolatitudes: Radically different climate system on the pre-Ediacaran Earth, 2016)
[16] (Betts, et al., 2018) Quote is from an Author Manuscript in PMC made available on February 20th, 2019.
[17] (Knoll, Javaux, Hewitt, & Cohen, 2006, p. 1026)
[18] (Nichols & Wörheide, 2005) with citation to McCaffrey's work from 1994.
[19] (Genitsaris, Moustaka-Gouni, & Kormas, 2011, p. 146) See Table 2.
[20] (Kumar, Stecher, Suleski, & Hedges, 2017) See data at http://www.timetree.org/api/widget/citation_data/2228
[21] (Seilacher, Bose, & Pflu¨ger, 1998, pp. 80, 82) On page 82, the presence of a coelom (a body cavity) was suggested by the burrowing movement of the corresponding animals.
[22] (Su, et al., 2021) An advance article.
[23] (NASA, 2019) The NASA Official is specified as Kristen Erickson for this part of NASA's website.
[24] Physicist Dr. Bernard Leikind, through our mutual friend physicist Dr. Art West, indicated via email that the evening leads according to the text. Dr. Leikind also wrote, "...that in ancient and in modern Jewish practice, a day begins at sunset" and that "...Today, for example, observant Jews begin their Sabbath on Friday night at sunset. The period of Sabbath rest lasts until Saturday at sunset." Dr. Leikind was and is correct on both counts.
[25] (Eldredge, 2002, p. 406)
[26] (Goudie & Cuff, 2002, p. 291)
[27] (Cornell, 1993, p. 2455)
[28] (Mayer, 2014, p. 57)

[29] (Schlager, 2008, p. 5)
[30] (Barton, Briggs, & Eisen, 2007, p. 73)
[31] (www.researchmedia.eu, 2014, p. 105)
[32] (van den Heuval, 2016) Abstract
[33] (Calvin, 1965, p. 15)
[34] (Cattermole & Moore, 1985, p. 12)
[35] (Burley, 2013, p. 100)
[36] (Spielman & Moore, 2020) Under the 'Methods' section
[37] (Boulton, 1987, p. 26)
[38] (Sinha & Tilton, 1973, p. 1845)
[39] (Dalrymple, 2004, p. 209)
[40] (Tatsumoto, Nunes, & Unruh, Early History of the Moon: Implications of U-Th-Pb and Rb-Sr Systematics, 1977, p. 507)
[41] (Tatsumoto, Nunes, & Unruh, Early History of the Moon: Implications of U-Th-Pb and Rb-Sr Systematics, 1977, p. 518)
[42] (Tatsumoto, Hedge, Doe, & Unruh, 1972, p. 1531)
[43] With the 4.7 billion years, I am referring to their uranium to lead and thorium to lead model from surface material. Within the source (Gopalan, Kaushal, & Wetherill, 1970), the authors also remark on a different age (4.75 ± .05 calculated from the ratio of $^{207}Pb/^{206}Pb$.)
[44] (Gopalan, Kaushal, & Wetherill, 1970, pp. 471-472)
[45] (Papanastassiou & Wasserburg, 1971, p. 43) See Papanastassiou and G.J. Wasserburg's Table 3 Apollo 14 ages.
[46] (Silver, 1970, p. 468) The Abstract

[47] (Tatsumoto & Rosholt, Age of the Moon: An Isotopic Study of Uranium-Thorium-Lead Systematics of Lunar Samples, 1970, p. 463) On page 463 of this source, the authors showed considerable uncertainty for the lower of the limit of their data and set it to 4.50 billion years ago. If a value on the lower bound of 4.50 billion years were to occur, this would result in the Sun coming in before the Earth: "We set the upper limit from our experimental error, which amounts to about 1.5 percent, and the lower limit to include Patterson's meteorite data."
[48] (Tatsumoto & Rosholt, Age of the Moon: An Isotopic Study of Uranium-Thorium-Lead Systematics of Lunar Samples, 1970, p. 461) From page 461, the full quote is "The fine dust and the breccia have a concordant age of 4.66 billion years on the basis of $^{207}Pb/^{206}Pb$, $^{206}Pb/^{238}U$, $^{207}Pb/^{235}U$, and $^{208}Pb/^{232}Th$ ratios. This age is comparable with the age of meteorites and with the age generally accepted for the earth."

[49] (Lowman Jr., Paul D (NASA Goddard Space Flight Center Greenbelt, MD, United States), 1971, pp. iii, 13) This is noted to have been published in the Journal of Geology, March, 1972, V. 80, No. 2, p. 125-166.
[50] (Ridley, 1975, p. 16) Immediately before stating the portion I quoted, in the preceding paragraph, Ridley stated that, "In order to understand the petrogenesis of the Moon, it is best to present the broad picture that has emerged since 1969, and later to follow up with more specific evidence. Although many important petrogenetic questions remain, the first-order evolutionary sequence has generally been accepted."(page 16)

[51] (Hays & Walker, 1977, pp. 132-133) This source had some reservations and considered the model 'highly speculative': "Despite these reservations, we present highly speculative model compatible with studies of lunar igneous rocks."
[52] (Cleeves, et al., 2014, p. 8)
[53] (Wall, 2014)
[54] (Engelking, 2014)
[55] (Fleur, 2016) a version of the article appeared on page D2 on April 26, 2016 in print in the New York Times according to website.
[56] (Cleeves, et al., 2014, p. 1)
[57] (Cleeves, et al., 2014, p. 2)
[58] (Bouvier & Wadhwa, 2010, p. 3)
[59] (Bouvier & Wadhwa, 2010)
[60] (Amelin, et al., 2010, pp. 343, 349)
[61] (Dunford, 2020)
[62] (Dalrymple, 2004, p. 179)
[63] ((auto-generated), n.d.) Termwiki
[64] (Yoshizaki, et al., 2019, p. 2) from the Introduction section. This is a preprint of an article that appeared later with different page numbering in the journal Geochimica et Cosmochimica Acta, Volume 252, 1 May 2019, pages 39-60.
[65] (Liu, Zartman, Ireland, & Sun, 2019, p. 18854)
[66] Stephen Mojzsis's website is http://isotope.colorado.edu/~mojzsis/ . The article is under 1997 for him, and he called the paper by a slightly different title at the time of this book's writing.
[67] (Nutman, Mojzsis, & Friend, 1997)
[68] A rounding of significant figures argument could be made, but with the description, it would be inaccurate to not describe this as a match given the information presented.
[69] dictionary.com's definition of a paleosol.
[70] (Britannica, The Editors of Encyclopaedia, 2017)
[71] (Eriksson, Mazumder, Catuneanu, Bumby, & Ilondo, 2006, pp. 182, 194)
[72] (Mukhopadhyay, et al., 2015) This age range is stated multiple times: in the abstract and in the section under the 'AGE OF THE KEONJHAR PALEOSOL' heading.
[73] (Eriksson, Mazumder, Catuneanu, Bumby, & Ilondo, 2006, pp. 169, 192)
[74] (Eriksson, Mazumder, Catuneanu, Bumby, & Ilondo, 2006, pp. 170, 192)
[75] The Nsuze Group was also believed by the scientists to have been above the waterline, but they express confidence in the rocks of the Dominion Group being above the waterline.
[76] (Eriksson, Mazumder, Catuneanu, Bumby, & Ilondo, 2006, pp. 169, 182, 192, 194)
[77] Consulted with a person who would prefer to remain anonymous.
[78] See Appendix B Supplementary Figure 4.
[79] (Freeman, 2005, p. 600)
[80] (Noffke, Christian, Wacey, & Hazen, 2013, pp. 1103,1118,1122)
[81] https://www.dictionary.com/browse/chlor-
[82] Accessed on March 14th, 2021.
[83] (Spores | plantbot, n.d.)
[84] (Galperin, 2013, p. 1)
[85] (Battistuzzi, Feijao, & Hedges, 2004) Under 'The colonization of land' heading, Battistuzzi refer Node H in Table I, which comes with a very large confidence interval: 2.70–3.49 billion

years ago by their authors remark, which they presumably rounded to 3 significant figures. Their table 1 data uses a 4th significant figure.
[86] (The Editors of Encyclopaedia Britannica, n.d.)
[87] (H6086 - `ets - Strong's Hebrew Lexicon (WLC).) See Appendix B Supplementary Figure 6.
[88] (H6086 - `ets - Strong's Hebrew Lexicon (WLC).) See Appendix B Supplementary Figure 5.
[89] (Voelz & Reichenbach, 1969, p. 856)
[90] (Kumar, Stecher, Suleski, & Hedges, 2017)
[91] (Tang & Joyce, 2021)
[92] (Gen 1 | ESV | STEP | In the beginning, God created the heavens and the ..., 2021) If one uses the ESV version on the stepbible.org website and if one hovers over the word 'made' in Genesis 1:16 for the phrase 'And God made the two great lights –…", a pop-up lists possible meanings of 'a.sah' for different verb formations. Under the Qal verb formation, meaning 1a2h) is 'to appoint, ordain, institute'. I use this meaning for a.sah / ʿāśâ in my rendering of the word. JUST ADDED
[93] Definition of sidereal from Oxford Languages online search.
[94] (Williams, Geological Constraints on the Precambrian History of Earth's Rotation and the Moon's Orbit, 2000, p. 50) This is recorded in his Table 1.
[95] Definition of a varve is from Definitions of Oxford Languages from online
[96] (Williams, Schmidt, & Young, Strongly seasonal Proterozoic glacial climate in low palaeolatitudes: Radically different climate system on the pre-Ediacaran Earth, 2016, p. 560)
[97] (Huynh, 2019)
[98] (Bond & Harris, 1988) Abstract
[99] (The Messianic Jewish Family Bible Society, 2015)

[100] (H3671 - kānāp̄ - Strong's Hebrew Lexicon (kjv), n.d.)
[101] (H3671 - kānāp̄ - Strong's Hebrew Lexicon (kjv), n.d.) See "The KJV translates Strong's H3671 in the following manner:" and "Strong's Definitions".
[102] (University of Wisconsin - La Crosse: Gillis, Rick; Haro, Roger J. , 2021)
[103] (Schlichting Jr., 1969, p. 951)
[104] (Smith, 1973, p. 876)
[105] (Smith, 1973, p. 878)
[106] (Diaz & Laybourn-Parry, 2019)
[107] (Genitsaris, Moustaka-Gouni, & Kormas, 2011, p. 146) See Table 2.
[108] (Moore, 1993) Dr. Joseph F. Saunders's obituary discusses his positions.
[109] (National Aeronautics and Space Administration, 1971, p. 292) See Table 1 for the Amoeba growth rates during flight and simulated flight.
[110] (National Aeronautics and Space Administration, 1971, p. 273)
[111] (National Aeronautics and Space Administration, 1971, p. 305)
[112] (Kumar, Stecher, Suleski, & Hedges, 2017) See data at http://www.timetree.org/api/widget/citation_data/2228
[113] (Kumar, Stecher, Suleski, & Hedges, 2017) Put the term 'Amoebozoa' in the 'Specify a Group of Taxa' text field, then press enter. Afterwards, under 'Rank', there is a 'Show Timetree' button. Click the button, then click on the root node (a black circle to the left), and it displays a time of "1522 MYA", which is equal to 1.522 billion years ago.

[114] (Genitsaris, Moustaka-Gouni, & Kormas, 2011, p. 146) See Table 2.
[115] (Kumar, Stecher, Suleski, & Hedges, 2017) See data at http://www.timetree.org/api/widget/citation_data/2228
[116] Although flatworms are also triploblastic, they lack a coelom. Speaking of flatworms, the beginning of the Genesis Day Three talks of 'the living creature', which may leave an attentive reader puzzled and asking, "Aren't all creatures living?" This reference to 'the living creature' may be a reference to land planarians: Geoplanidae. Scientists from the University of Nottingham in the UK have remarked that planarians are notable for limitless capacity to regenerate aged cells, apparently rendering them unable to die from aging: "Planarian worms have amazed scientists with their apparently limitless ability to regenerate. Researchers have been studying their ability to replace aged or damaged tissues and cells in a bid to understand the mechanisms underlying their longevity." (University of Nottingham, 2012) The write-up cites the journal article: Thomas C. J. Tan, Ruman Rahman, Farah Jaber-Hijazi, Daniel A. Felix, Chen Chen, Edward J. Louis, and Aziz Aboobaker. **Telomere maintenance and telomerase activity are differentially regulated in asexual and sexual worms**. *Proceedings of the National Academy of Sciences*, February 27, 2012 DOI: 10.1073/pnas.1118885109
[117] (Genitsaris, Moustaka-Gouni, & Kormas, 2011, p. 146) See Table 2.
[118] (Kumar, Stecher, Suleski, & Hedges, 2017) See data at http://www.timetree.org/api/widget/citation_data/2228
[119] (29.2B: Gnathostomes - Jawed Fishes - Biology LibreTexts, 2021) Under '29.2B: Gnathostomes - Jawed Fishes'
[120] (Peterson, 1994) Abstract
[121] (Genitsaris, Moustaka-Gouni, & Kormas, 2011, p. 146) See Table 2.
[122] (Kumar, Stecher, Suleski, & Hedges, 2017) See data at http://www.timetree.org/api/widget/citation_data/2228
[123] (Eicher, 2020)
[124] (Terlevich & Forbes, 2002) A picture of their Table 3 data is available at https://academic.oup.com/view-large/figure/92737888/330-3-547-tbl005.jpeg
[125] (Forbes, Ponman, & Brown, 1998)
[126] (Rampazzo, et al., 2007) The article is stated to have appeared in the October 2007 issue #1.
[127] (Schweitzer & Seitzer, 1998, p. 2206) in the Abstract.
[128] (Schweitzer & Seitzer, 1998, p. 2207)
[129] (NASA, 2019) The NASA Official is specified as Kristen Erickson for this part of NASA's website.
[130] It is possible that the 500 million year estimate was originally derived from an overly low estimated of age galaxy I Zwicky 18, but since I Zwicky 18 was found to be older since at least 2007, that galaxy is not relevant.
[131] (Cain, 2004)
[132] (Genitsaris, Moustaka-Gouni, & Kormas, 2011, p. 146) See Table 2.
[133] (Kumar, Stecher, Suleski, & Hedges, 2017) See data at http://www.timetree.org/api/widget/citation_data/2228
[134] (NASA Official: Dr. Edward J. Wollack, n.d.)
[135] (How Old Are Galaxies? | NASA Space Place – NASA Science for Kids, 2019) The NASA Official responsible for the site is Kristen Erickson, and the Program Manager is Heather Doyle.

[136] (Mosley Jr., 2021) PRESERVED
[137] (Hocken, n.d.)
[138] (Husfeld, 1997)
[139] (Bikos & Jones, n.d.) I also accessed Wikipedia on February 21st, 2022.
[140] (Hocken, n.d.) This information appears under the section titled 'New Leap Year Formula'.
[141] (NASA Glenn Research Center, n.d.)
[142] (H1877 - deshe' - Strong's Hebrew Lexicon (WLC).)
[143] (H6086 - `ets - Strong's Hebrew Lexicon (WLC).)
[144] (NASA, 2016)

About the Author

Kevin Bradford Ornellas is an IT specialist who resides in Honolulu, Hawaii. Kevin can be reached at kevinbradfordornellas@gmail.com. Check-out the Science and God Alliance (SAGA) Facebook groups.

Made in the USA
Monee, IL
04 November 2026

67385486R00105